SpringerWienNewYork

Peter Weibel

Garnet Ord

Otto E. Rössler

Editors

Space Time Physics and Fractality

Festschrift in honour of Mohamed El Naschie
on the occasion of his 60th birthday

SpringerWienNewYork

Editors:
Univ. Prof. Peter Weibel
Center for Art and Media, Lorenzstr. 19, D-76135 Karlsruhe, Germany

Prof. Garnet N. Ord (PhD)
Dept. of Mathematics, Ryerson University, Toronto, Ont. Canada., M5B2K3

Prof. Dr. Otto E. Rössler
Dept. of Theoretical Chemistry, University of Tübingen, Auf der Morgenstelle 8,
D-72076 Tübingen, Germany

A *ZKM | Institute for Basic Research* edition

The papers have been presented in the symposium "Space Time Physics, Transfinite Mathematics and Computer Art", ZKM – Center for Art and Media, Karlsruhe, 11 October 2003, organized by H.H. Diebner and P. Weibel.

This work is subject to copyright. All rights reserved, whether the whole or part of the material is concerned, specifically those of translation, reprinting, re-use of illustrations, broadcasting, reproduction by photocopying machines or similar means, and storage in data banks.

Product Liability: The use of registered names, trademarks, etc. in this publication does not imply, even in the absence of specific statement, that such names are exempt from the relevant protective laws and regulations and therefore free for general use.

© 2005 Springer-Verlag/Wien
© 2005 by the authors
The authors are responsible for their contents
Printed in Austria
SpringerWienNewYork is a part of Springer Science + Business Media
springeronline.com

Cover Design: Holger Jost
Printed by: Druckerei Theiss GmbH, 9431 St. Stefan, Austria
Printed on acid-free and chlorine-free bleached paper
SPIN: 11403722
With numerous figures
Library of Congress Control Number: 2005928383
ISBN-10 3-211-25210-X SpringerWienNewYork
ISBN-13 978-3-211-25210-9 SpringerWienNewYork

Contents

Space Time Physics and Fractality

Editors:
Foreword ... VII

Amr Elnashai:
Recollections ... XII

Werner Martienssen:
Congratulations ... XV

Authors ... XVI

Paolo Grigolini:
Quantum Mechanics and Non-Ordinary Statistical Mechanics ... 1

Garnet N. Ord:
Bohr, Bohm and Entwined Paths ... 31

Laurent Nottale:
On the Transition from the Classical to the Quantum Regime in Fractal Space-Time Theory ... 41

G. Iovane:
Mohamed El Naschie's $\epsilon^{(\infty)}$ Cantorian space-time and its consequences in cosmology ... 53

Otto E. Rossler:
Needle People and Pancake People: The Gulliver Effect ... 60

David Ritz Finkelstein:
Cosmic Computation ... 65

Walter Greiner and Andrey Solov'yov:
Atomic Cluster Physics: New Challenges for Theory and Experiment   83

Helmut Kröger:
Why are Probabilistic Laws Governing Quantum Mechanics
and Neurobiology?   98

B.G. Sidharth:
Ramifications of Non Commutative Spacetime   135

Karl Svozil:
Computational Universes   144

Hans H. Diebner and Florian Grond:
Usability of Synchronization for Cognitive Modeling   174

Tomasz Kapitaniak:
Riddling Bifurcation and ... Interstellar Journeys   184

Reint de Boer:
Theoretical Poroelasticity – A New Approach   189

Mohamed S. El Naschie:
From Hilbert space to the number of Higgs particles
via the quantum two-slit experiment   223

Foreword

My first sighting of Mohamed was in a picturesque little village in Switzerland. We had agreed to meet in a parking lot, and as I sat in my car waiting, I noticed a little yellow jeep tearing down the mountain side at high speed. Assuming the driver to be a teenager, I was surprised when the jeep screeched to a halt beside me and a handsome and distinguished looking man leapt out and greeted me warmly. Grinning boyishly, Herr Professor addressed me as 'Fractal Man', chided me for taking so long to cross the alps to meet him, and bade me follow him up the hill to my hotel. In the ensuing high-speed chase I taxed the limits of my little car to keep Mohamed in sight. Twice I had to back rapidly out of a cul-de-sac to avoid being run over by my host, who was clearly enjoying the drive, if not entirely certain of where he was going. The route eventually revisited the parking lot we had started from, and by the time we reached the hotel I had absorbed the charms and ambiance of the little village, having traversed most of its streets and alley ways at least once. When the dust had settled and I was safely installed in my hotel, my genial host invited me over to discuss physics. It was a fantastic evening that I shall never forget. My own work was not well known and was very much outside the mainstream. I had travelled my own path for fifteen years and considered I knew every rock, pebble and fallen branch on the route I had taken. I was also familiar with trying to entice colleagues into the uncharted territory that my trail explored. It was usually a painful and unrewarding process. Physicists tend to be conservative, preferring the tidy well-kept routes of a civilized hierarchy. Anticipating the usual reluctance, I was surprised that my mischievious friend not only offered no resistance to exploring my path, he leaped over me like a gazelle and took off along the route in what I thought to be indecent haste. In vain I tried to slow him down, pointing out this treasured rock, that carefully assembled marker. But, as all who are lucky enough to have discussed physics with Mohamed will know, resistance is futile. Listening to Mohamed gambol along what I considered 'my path' was a revelation indeed. Mohamed never ambles, he leaps. The added height of those leaps allows him to see many things hidden to the average traveller. Although I knew and treasured the local details that Mohamed would cross in a single bound, I had never seen the surrounding country that he was able to see. Mohamed embedded my one-dimensional path in a landscape of higher dimension, and more intricate beauty. That landscape was riddled with the ideas and paths of many people, all woven into a coherent fabric by a gifted raconteur. Like the trip up the mountain to the hotel, I was thoroughly stretched to follow his enthusiastic pace, but in the process I saw the beauty of an inspired synthesis of ideas that would have otherwise been beyond my vision. Science is, in the end, a process of exploration. Few people have the combined talents and circumstances to explore more than a small patch of well-charted ground. One needs the intuition that is the scientist's compass, the knowledge and imagination that is his spy-glass, the technical skill to travel swiftly and safely, and finally the courage to visit, alone if necessary, territory that is overlooked or spurned by colleagues. Mohamed has all of these qualities in larger-than-life measure.

This special issue is from a few friends who rejoice in, and benefit from the inspiration of Mohamed's passion for Physics. It is a small but affectionate salute to our favourite Editor/Explorer-in-chief, on the occasion of his 60th birthday.

Garnet Ord
Toronto
October 5, 2003

Cracking the Enigma of Space-Time

We all live in the yellow submarine – in the age of the telematic revolution. This is also the age in which science and art come close together again, closer than at any other epoch since pre-paleolithic times. This time around, the unified view is no longer reserved to the caste of shamans and berdachs, however, but is perceived by everybody to be an inalienable right. The re-unification lies right around the corner. Gödel, Turing, the computer and the DVD have created a new reality and riddle at the same time: To see ourselves imprinted into the fabric of the brick wall, like a human shadow burned into the ground after an atomic flash – Gödel's frightening insight. It deserves to be elucidated further as long as it is still fresh in our minds. This book brings together in one place some of the most dedicated workers in this new no-man's and nomad's land, all gathered to honor one of the most prolific minds in the Gödelian realm of the transfinite – Mohamed El Naschie. The symposium held at the Karlsruhe Media Lab in October 2003 paid tribute to the "cosmic computation" (David Finkelstein) initiated by this multifaceted fractal mind. His closest co-voyagers, Garnet Ord and Laurent Nottale, share with him the conviction that a path-integral-like view of the quantum enigma holds the key to a deeper understanding of the cosmos. The mark imposed by this approach on high-energy physics it was El Naschie's privilege to discover. An infinite-dimensional Cantor dust is – according to this view – the real fabric of nature. Being part of this "dust" (which despite its infinite dimensionality retains some biblical connotations) *contracts* you into becoming, in your own eyes, a three-dimensional body made up of particles in a 4-dimensional space-time. Carl-Friedrich von Weizsäcker's early challenge to explain why space has three dimensions and time has one, encouraged by Pauli and Weyl, has come closer to its resolution to date in the hands of this scientist. Not only the visible (and invisible string-theoretic) dimensionalities of space become deducible; even the fine structure constant itself and also the muon-electron mass ratio (to mention only one among many) can for the first time be predicted with infinite precision. "Thinking transfinite" in the spirit of Mandelbrot and the pre-Socratic philosopher Anaxagoras, the inventor of chaos theory who was not far from the ancient Egyptian tradition, proves amazingly successful to date. The only question not yet addressed in this book of laudatory contributions is: why is this so? We have no solid ground under our feet as is well known. There is almost nothing but emptiness – as if one were looking down from an orbiting spaceship all alone. Why does the fractal dust qualify as so solid a basis despite the fact that it is virtually nothing but holes, holes within holes, and so in infinitely many dimensions? We – the present introductory writers – have asked this question to Mohamed. Two possible answers took shape in the 3-D sounding board situation between the three of us, punctured by his laughter. Both are not mutually exclusive. The first is "faith." Clear thinking in physics (and elsewhere) is not possible without an absolute faith in a benevolence-beyond that is the hallmark of religion, as Einstein never tired to repeat in one way or the other. The real riddle is "assignment" – someone (who?) is being given a place in a universe of consciousness and, within it, a universe of structures – dust as far as we can now tell. There is no water to swim in, in this medium. Science itself is only concerned with the least lively, shadow-like aspect – the Cartesian relations. All that we have to cling to is an interface (and the latter exists only for one moment, the now, which may turn out to be indestructible). When carefully looking through the meshes, we find other meshes in an infinite-dimensional continuum of holes, as it were. This is the quantum reality which now turns out to cover, not just the states of molecules and nuclei, but also everything below that level, even though nothing was thought to exist below that level until quite recently. The string flute itself is made of dust! This reality at the same time extends to the largest structures in the universe: There is nothing but voids beyond voids in all directions, both upstairs and downstairs (the scale relativity of Nottale). The path integral is weaving together lumps and dimensions and strings: Feynman's many paths extend the submicroscopic to the heavens. This unusual vision, borne out of faith in the determining force of the fragrance-giving instance, is the organizing center of this book. But there is a second insight arrived at in the above-mentioned sounding board situation. If the quantum riddle is so pervasive that even the rigid curtains of space and time are fractured and fractalized into infinitely many pieces and dimensions: could it be that this very fact can be explained once more on a further level? While many of El Naschie's string-theorist followers would be reluctant to seriously ponder the existence of an even more fundamental (exo) level of nature, the originator himself is not afraid of such a scenario. The interface view – the media-theoretic paradigm – likewise calls for the presence of a more outside level. Good old Boltzmannian rationalism – classical determinism – may (just may) still be lurking behind the infinitely complexified quantum world. While the great successes of Mohamed's let the desire for such a rationalist picture, in the footsteps of 18th century physicist and Jesuit priest Roger Joseph Boscovich, appear outdated at first sight, the revival of the interface idea in our current age of the computer game and telepresence (even after death) calls for a synthesis. There is always a hidden level – as the present book itself is living proof of. The dust of letters contained in it owes its existence to a hidden level: Lydia and Reimara inaugarated it and Anke made the symposium possible. Science is friendship just as art is friendship. Every user of an art museum pretends he or she does not see it *is* a joke. But everybody knows that art is a joke. Just as the smile of the toddler is, when he returns that of Mom's with his own unique mischievous grin. But then there is a second level on which the mutual pretense is transformed into genuine closeness of an infinitely reliable kind. Art is a smile that can be relied upon. The world, too, can smile, and so in infinitely many dimensions. Let us see whether the smile of Mohamed comes through in the splintered ramified reflections that await you in the following pages.

Peter Weibel and Otto E. Rössler
Karlsruhe
November 6, 2004

Symposium Announcement

On the occasion of the 60th birthday of Mohamed El Naschie the ZKM is organizing a symposium on "Space-Time Physics, Transfinite Mathematics, and Computer Art". Mohamed El Naschie was successful in deriving a model for the so-called multifractal space-time which allows to predict the mass of elementary particles and derive universal constants. The derivation of these constants is not possible through the "standard model" which is why the physicists search for extensions or modifications of this theory. The conference deals with problems of this kind and includes thereby new and controversially debated approaches. It also includes mathematical considerations that propose them-selves to be fruitfully used within space-time physics. Additionally, the symposium addresses the attempt to use artistic approaches to gain knowledge on space-time and it is dicussed which connections between physics and art exist.

Hans Diebner, Peter Weibel, Otto E. Rössler
Karlsruhe, October 2003

Recollections

by Amr Elnashai

On his 60th birthday, Mohamed El Naschie (or El Nashie, or Elnashie, or Elnashai, depending on whom you believe amongst the three brothers and their father regarding how each have decided to spell his peculiar surname in English) is as productive, prolific and vigorous as anyone on God's earth. After considerable pondering, I decided to offer this short article in lieu of a technical paper. The reasons are many, some are a cause of constant argument between Mohammed and I, in the most congenial manner possible between two El Naschies, or El Nashaies or Elnashies or, finally, Elnashais (the latter is of course the right spelling!!). First and foremost, the only subject which I can write about authoritatively, and some would consider this to be a stretch, is earthquake engineering. If I do so, I will have to endure a length and well-rehearsed, and oft-received, lecture on how narrow I am, having spent more than half a life-time researching earthquake response of structures and exploring the intriguing relationship between ground motion characteristics and response of complex structural systems. I have not cracked it yet. But my brother believes that I suffer from total lack of imagination and technological flatness because I do not change disciplines every ten years. A colleague of mine just the other day criticized a young faculty recruit because he wrote a paper on scaling of earthquake records for structural analysis purposes, because he 'has not been trained in engineering seismology and related fields'. I would love to lock up the said colleague with Professor El Naschie, or ..., and reveal to the former that the latter was trained as a structural engineer, and is now a pinnacle of nuclear and particle physics, amongst other things. So, once again, one man's terrorist is the other man's freedom fighter. To my dear UIUC (this is the acronym of my beloved employer, the University of Illinois at God-Forsaken Urbana-Champaign, the best technical institution that there is, located on the most boring spot in the Universe, and beyond) colleague, I have demonstrated clear signs of insanity, coupled with recklessness and lack of appreciation of the intricacies of science and technology, and disciplinary boundaries, merely because I cross-publish!! To my dear brother Mohammed, I am conservative, unimaginative, lack technical confidence and certainly missed on audacity when they were distributing some.

He, Mohammed, is indeed brilliant, but all those who will read my article know that already. He is versatile and is a visionary, but these are characteristics that all those who meet him conclude that he enjoys. So, what is new from his kid brother? My recollections of him when I was about 8 years old are numerous and rather entertaining. He is the best narrator of films

I know of. The films he narrated to me stick in my mind and soon after I cannot say for sure if I listened to him narrating the film or did I see it in the theatre? My son, Shadie, thinks I am good at telling tales. Well, I am but a drop in Mohammed' ocean of impressions, background ambiance and ability to whisk the listener to where the action took place, virtually. I also recall him as rather eccentric, strict and very funny. As I grew older, I grew closer to our middle brother who, notwithstanding our current predicament, was a terrific older brother. I continued though to say that although I do not see Mohammed much, I admire him from a distance; whilst I could not afford the same admiration to our middle brother Saiid, who really looked after me and helped me grow up. At the age of 12 I went to Germany and spent Christmas and New Year with Mohammed at 10 Schmide Strasse, Hannover, Germany. The three weeks were to change my life and outlook. I will not go into what happened in Hannover so many years ago lest this volume be censored for violation of British decency laws. Suffices it to say that I decided to stay in Germany. When Mohammed mentioned this to our father on the phone, the latter threatened to contact the Egyptian Consulate and bring me back by force! He is kind, our father, he did not know what awaited him from the third in line of the El Naschie trio. I went back after a truly wild three weeks in Hannover, including a hair-raising Carnival night. Visiting my brother in Germany was 'the end of innocence'.

When I was 17, I visited Mohammed in London, and followed this by a visit to Saiid in Edinburgh. There was a hot competition between them regarding who would be able to convince me to leave the Medical School at Cairo University and go into engineering; civil, like Mohammed, or chemical, like Saiid. Mohammed did not offer other than advice. Saiid offered restocking my racing pigeons lofts from top Highlands breeders. I took Saiid's offer, but followed Mohammed's advice, and transferred my registration from Medical School to Civil Engineering, three weeks into the academic year. Another visit to Mohammed at his South London apartment in Streatham Hills had a another significant effect; I decided to pursue an academic career. This was in 1973, the year my name appeared on a paper, with Mohammed, published in the Journal of the American Institute of Aeronautics and Astronautics. It remains as the first journal paper on my curriculum vitae, notwithstanding that I do not understand a word in it. I followed in the footsteps of my eldest brother and pursued research in advanced structural engineering. In time, he cheated and moved to physics. At that juncture, we parted technical company; I could not possibly take the chances that he has taken, from the top of the structural engineering food-chain to a high-flying and accomplished physicist! I made excursions in strong-motion data analysis, but they were shy and terribly cautious, at least from my point of view. I gradually learned that Mohammed's abilities are not normal. He is a full-blown genius. I decided that my best strategy is to stick to main-stream earthquake engineering, and claim sibling credit for my brother's extra-ordinary achievements.

Being Mohammed El Naschie's brother is a major responsibility! You are expected to be somewhat of a genius, which I certainly am not. OK, I still hold the dubious distinction of being the youngest full professor in Civil Engineering at Imperial College, London, and certainly the first, and hitherto the only, Middle Eastern full professor and Head of Section, but how high this stacks up next to the series of professor positions that Mohammed has piled up in four continents? As his brother, he expects extremely accurate and verifiable statements, even in assessing a restaurant or a new film! If one can attract praise from Mohammed, one has reached the Promise Land. Until this day, I am inspired along the route of 'publish or perish' by Mohammed's publication record (journal papers in the hundreds) and the story of Thomas Harriot. To those who do not know who he is; the majority of people, Harriot discovered the telescope months before Galileo Galili of Padova in 1609. But old Harriot did not publish the paper, only wrote a few comments in his private notebook. Galili, on the other hand, almost had the paper written, with gaps left for the results! The telescope is widely attributed to Galili, because of timely publication I also learned from Mohammed to look beyond the obvious, and to make as few assumptions as possible, or at least this is what I thought he advised me to do! As I approach the big Five-O, I still seek Mohammed's advice, much to his dismay, because once he knows about a problem I am facing, he cannot walk away from it; this is his nature. Did I say complimentary things about my brother? I do not know, do not care, and I am sure neither does he. What I am sure of is that I said what I feel about him. Mohammed's re-emergence in my life around the year 2000, was one of the most profound events in the past 49 years. It took us about 20 years to work out that we are both well-meaning and that we love each other much. So, Happy Birthday my dear brother, and may we both celebrate your 80th birthday together.

Amr Elnashai (the correct spelling!)

Dear Mohamed,

my warmest congratulations to your 60th birthday. This is a great day for you and for all those who think with you and who feel with you. On the 10th of October our thoughts will be with you; our wishes: Keep your very good health, stay on your way, continue to give free rein to your fascinating ideas, we need you! Mohamed, in short, for the next 60 years: Let us keep our friendship!

Together with my coworker Bernd Hils I wanted to prepare an experiment for your birthday. It is not ready in the moment, but we will work on it, and I will show it to you, when you come next time to visit our lab. Garnet very kindly offered me that we can still publish it in a later issue.

It's on Berry's phase. You know, Berry's phase is one of those phenomena which are considered as a typical quantum-physical subject, nevertheless it played a roll in experimental optics already before quantum physics was discovered; so, in a way, it is a bridge between classical and quantum physics. I am very interested to learn your opinion.

Looking forward to experience a great birthday,
with all my good wishes
sincerely yours
Werner

Authors

REINT DE BOER, Institut für Mechanik, FB 10 – Bauwesen, Universität Essen, 45117 Essen, Germany.

HANS H. DIEBNER is currently head of the Institute for Basic Research at the Center for Art and Media, Karlsruhe, Germany. He is trained in physics and received his doctoral degree in 1999 for his work on exactly reversible molecular dynamics simulations. He was supervised by Otto Rössler who taught him both producing chaos and to control it. During his post-doctoral position in the Institute for Medical Biometry in Tübingen Diebner applied his system theoretical skills to immuno- epidemiological modeling especially in the case of malaria tropica. He currently focuses on the modeling of cognitive systems and tries to bridge natural sciences, philosophy and arts.
Editor:
Hans H. Diebner and Lehan Ramsay (Eds.), Hierarchies of Communication. Center for Art and Media, Karlsruhe 2003.
Hans H. Diebner (Ed.), Studium generale zur Komplexität. Genista-Verlag, Tübigen 2001.
Hans H. Diebner, Timothy Druckrey, Peter Weibel (Eds.), Sciences of the Interface. Genista-Verlag, Tübingen 2001.
Address: Center for Art and Media, Lorenzstr. 19, 76135 Karlsruhe, Germany. www.diebner.de E-mail: hd@zkm.de

AMR ELNASHAI, Urbana, Illinois.

MOHAMED EL NASCHIE, Frankfurt Institute of Advanced Studies, Frankfurt, Germany and Dept. of Physics, University of Alexandria, Alexandria, Egypt.

DAVID RITZ FINKELSTEIN teaches and studies physics at Georgia Institute of Technology and edits the International Journal of Theoretical Physics. His main work is to extend quantum logic to still deeper levels of physics. As by-products of this main interest, he contributed to early work on the topology of the gravitational field, the concept of the black hole, the gauge theory of the electroweak interactions, and quantum theory. He is also working on an analysis of Albrecht Dürer's engraving *Melencolia I*. His works in progress are posted at http://www.physics.gatech.edu/people/faculty/dfinkelstein.html
Address: School of Physics, Georgia Institute of Technology, Atlanta.

WALTER GREINER, Prof. Dr. Dr. h.c. mult. of the Goethe University, Frankfurt am Main has made outstanding contributions to nuclear physics in Europe and world-wide over the past 4 decades. He has initiated and developed the theoretical understanding of a wide range of phenomena in nuclear physics, from nuclear structure (Rot-Vib model, Gneuss-Greiner model, giant resonances) and nuclear reactions, to the theory of strong quantum fields. His pioneering theoretical research has stimulated many innovative ideas in the experimental and theoretical nuclear physics community.

He is one of the fathers of the highly successful German laboratory for basic research in heavy ion physics, GSI in Darmstadt. His ideas ('cold fusion valleys' and fragmentation theory) have driven the successful experimental search for very heavy nuclei (Z=107-112) from the early beginnings in the mid sixties to now.

He has pioneered and steered the development of the field of relativistic heavy ion collisions since 1973, which broke the ground for the physics justification and realisation of major experimental facilities world-wide, at the Berkeley BEVALAC, GSI's SIS, RHIC at Brookhaven National Lab.

His theoretical predictions have driven many of the large scale experimental collaborations to work on the formation of ultra dense, hot matter, resonance matter and strange matter.

Walter Greiner's early suggestions to study the properties of the nuclear equation of state via collective flow (bounce-off and squeeze-out) and his subsequent developments of macroscopic and microscopic theories have brought the breakthrough to this field and to the resulting search for quark matter and exotic states of matter at CERN's SPS and the future LHC-accelerator.

During the last decade, Walter Greiner has made a number of important contributions to atomic cluster physics, which has numerous parallels with nuclear physics.

His internationally renowned textbooks, in particular the three volume monograph on 'Theoretical Nuclear Physics' with the late Judah M. Eisenberg (Tel Aviv), with Elsevier, and the 17 volume series on 'Theoretical Physics' with Springer Verlag and Harri Deutsch Verlag, which appeared in 7 languages, have attracted generations of young students and postdocs into theoretical physics, world-wide.

Address: Institute for Theoretical Physics, Robert Mayer St. 10, D-60054, Frankfurt am Main, Germany, e-mail: greiner@th.physik.uni-frankfurt.de

GERARDO IOVANE, Dipartimento di Ingegneria dell'Informazione e Matematica Applicata, Universitá di Salerno, Italy, e-mail: iovane@diima.unisa.it

PAOLO GRIGOLINI is sharing his research time between the Center for Nonlinear Science of the University of North Texas and the Interdisciplinary Center for Complexity of the University of Pisa. The main goal of his research program is twofold. On one side, he has recently led a group of co-workers to the construction of an efficient technique of time series analysis, which detects the statistical properties of crucial events, either visible or invisible, i.e., either recorded or not. This technique of analysis is based on the dynamic approach to anomalous diffusion, renewal and subordination. On the other side, he is studying intermittent effects produced by new material such as blinking quantum dots, aiming at making ostensible the limitations of the ordinary quantum mechanical approaches.
Address: Center for Nonlinear Science, University of North Texas, PO Box 311427, Denton, TX 76203-1427

FLORIAN GROND was born in Graz in 1975. He studied Chemistry in Graz, Leicester and Tübingen with a focus on complex system theory. Since 2001 he works in Karlsruhe in the Institute for basic research at the ZKM. His main interest is in the field between science (nonlinear dynamical systems) and art with an alternating focus. Parts of his work were exhibited in Madrid and Karlsruhe and Graz.
Address: Center for Art and Media, Lorenzstr. 19, 76135 Karlsruhe, Germany. www.sol-sol.de E-mail: grond@zkm.de

TOMASZ KAPITANIAK's research covers wide range of different subjects of nonlinear dynamics as: Chaos, Stochastic Mechanics, Nonlinear Vibrations, Synchronization of Systems. He has authored 11 books, edited 4 conference proceedings and 8 journal special issues, 5 chapters in books, and over 120 papers and conference presentations. He acts as Editor-In-Chief of the Journal: Mechanics and Mechanical Engineering, Associate Editor of Chaos, Solitons and Fractals and as co-editor in two other journals.
Address: Division of Dynamics, Technical University of Lodz, Stefanowskiego 1/15, 90-924 Lodz, Poland.

HELMUT KRÖGER teaches physics at Laval University in Québec, Canada. He works in theoretical and computational physics in the domain of cosmology, condensed matter, quantum chaos, neural networks and computational neuroscience.
Address: Département de Physique, Université Laval, Québec, Québec G1K 7P4, Canada. E-mail: hkroger@phy.ulaval.ca

WERNER MARTIENSSEN, Physikalisches Institut der Universität Frankfurt am Main, Robert Mayer Strasse 2-4, D-60054 Frankfurt/M, Germany. E-mail: Martienssen@Physik.uni-frankfurt.de

LAURENT NOTTALE is currently a Directeur de Recherche at the Centre National de la Recherche Scientifique (France) and he works at the Paris Observatory. He contributed to works in extragalactic astrophysics and cosmology, studying in particular gravitational lensing. His main interest is now in the foundation and the development of the theory of fractal space-time and scale-relativity (which is based on non-differentiable geometry) and in its applications to various sciences, including physics, astrophysics and sciences of life. Author: L. Nottale, Fractal Space-Time and Microphysics: Toward a Theory of Scale Relativity (347 pp). World Scientific, Singapore 1993.
L. Nottale, L'Univers et la Lumiére: cosmologie et mirages gravitationnels (288 pp). Flammarion, Nouvelle Bibliothéque Scientifique, Paris 1994.
L. Nottale, La Relativité dans tous ses Etats : Au delà de l'Espace-Temps (319 pp). Hachette, Paris 1998.
L. Nottale, J. Chaline, P. Grou, Les arbres de l'évolution: Univers, Vie, Socéités. (379 pp.) Hachette, Paris 2000.
Address: LUTH, Observatoire de Paris-Meudon, F-92195, Meudon Cedex, France. URL: http://wwwusr.obspm.fr/~nottale E-mail: laurent.nottale@obspm.fr

GARNET N. ORD, Dept. of Mathematics, Ryerson University, Toronto, Ont. Canada., M5B2K3

OTTO E. ROSSLER, born 1940 in Berlin, studied medicine in Tübingen. After receiving his doctorate in 1966, Rossler was a postdoctoral fellow at the Max Planck Institute for Behavioral Physiology in Seewiesen. Since 1970 he is a professor at the University of Tübingen, where he teaches nonlinear dynamics, dissipative structures, chaos theory, mental equations, fundamental physics, endophysics, and deductive biology at the Institute for Physical and Theoretical Chemistry. He was a visiting professor at many international universities and held lectures in several countries. He has published over 300 scientific works on artificial life and artificial brains; on bifurcation, differentiable mechanisms, chaos attractants, hyperchaos, endophysics, micro-relativity, and computer interfaces, as well as works on Anaxagoras, Descartes, Lebniz, Boscovich and Kant – and a recent work on the Internet project "Lampsacus." Rossler co-edited numerous scientific journals. He wrote and co-authored numerous books such as: '"Endophysics: The World of the Internal Observer", "Jonas' World" (1994), "The Flaming Sword or How Hermetic is the Interface of Microconstructivism?" (1996).

Address: Division of Theoretical Chemistry, University of Tübingen, Auf der Morgenstelle 8, 72076 Tübingen, F.R.Germany

B.G. SIDHARTH, the Founder Director of the B.M. Birla Science Centre, Hyderabad, has been working on his model of Fuzzy Spacetime and fluctuations. The resultant cosmology correctly predicts a dark energy driven acceleration by observation. His work also points to the reconciliation of gravitation and electromagnetism and provides a mass spectrum for all known elementary particles. It not only deduces the otherwise empirically known so called large number relations, considering to be accidents but also gives a description of gravitation that explains the mystery of the Weinberg formula which links microphysical parameters to large scale parameters.
Address: Centre for Applicable Mathematics & Computer Sciences, B.M. Birla Science Centre, Adarsh Nagar, Hyderabad - 500 063 (India).

ANDREY SOLOV'YOV currently heads a group at the Frankfurt Institute for Advanced Studies and focuses on theoretical studies of complex molecular systems, such as atomic clusters (tiny pieces of matter consisting of just tens or hundreds of atoms), biomolecules (amino-acids, proteins, antibiotics), nanostructures deposited on a surface. He tries to find answers to wide-ranging questions in multi-atom systems, which concern the principles of matter self-organization, self-assembling and functioning on the nanoscale. The description of nonrelativistic or relativistic many-electron systems using quantum ab-initio and model approaches is another topic of his research. The understanding of structure and dynamics of mesoscopic systems like atomic clusters and macromolecules lies at the heart of a large variety of problems at the forefront of physics, chemistry and biology. Research conducted by Prof.Dr. A.V.Solov'yov contributes to a number of these fields.
He has developed a variety of new theoretical methods allowing detailed theoretical description of the electronic and ionic structure of complex multi-atomic systems, which have been utilized for studying metal and noble gas clusters, fullerenes and bio-molecules. Also, the dynamic properties of these systems manifesting themselves in collision processes as well as in fission and fusion processes have been investigated. He has predicted a new type of undulator radiation, the so-called crystalline undulator radiation, which is generated by a motion of charged particles through a periodically bent crystal. It was demonstrated that the crystalline undulator can be used as a new, powerful source of high-frequency monochromatic electromagnetic radiation of a free-electron laser type. He has developed the most advanced theory of the polarizational bremsstrahlung radiation emitted by many-electron systems, such as many-electron atoms, complex molecules and clusters, due to their polarization in collision processes.

He is an author and co-author of more that 150 journal publications and 6 books.
Books:
1. W. Greiner, A.V. Solov'yov, S. Misicu (eds.), Channeling-Bent Crystals-Radiation Processes, EP Systema Bt, Debrecen (Publisher), Hungary (2003), p. 1-234
2. A.V. Korol, A.G. Lyalin, A.V. Solov'yov, Polarizational bremsstrahlung, Publishing House of St. Petersburg State Polytechnical University, St. Petersburg (2004), p. 1-300 (in Russian)
3. J.-P. Connerade, A.V. Solov'yov (editors), Latest Advances in Atomic Clusters Collision: Fission, Fusion, Electron, Ion and Photon Impact, Imperial College Press, World Scientific, Coven Garten, London WC2H 9HE, UK (2004), p. 1-396
Address: Frankfurt Institute for Advanced Studies, Robert Mayer St. 10, D-60054, Frankfurt am Main, Germany, e-mail: solovyov@fias.uni-frankfurt.de
Address: A.F.Ioffe Physical-Technical Institute, Politechnicheskaya 26, 194021 St.Petersburg, Russia.

KARL SVOZIL, professor of physics at the TU Vienna, has worked in a number of research institutions; among them in the US at UC Berkeley & LBL, in Russia at the Lomonosov University Moscow & Lebedev Institute and more recently in New Zealand. He is associated editor of several research journals and external faculty member of the CDMTCS of The University of Auckland. His research activity comprises two monographs and over a hundred contributions to research journals. Currently his main interests are in quantum information and computation theory, as well as in coding strategies for nerve signals (stochastic interference), and in quantum and automaton logic. Author: K. Svozil, "Randomness and Undecidability in Physics" (World Scientific, Singapore, 1993), xvi+292 p. K. Svozil, "Quantum Logic" (Springer, Singapore, 1998), xviii+214 p.
URL: http://tph.tuwien.ac.at/~svozil E-mail:svozil@tuwien.ac.at
Address: Institute for Theoretical Physics, Technische Universität Wien, Wiedner Hauptstrasse 8-10/136, A-1040 Vienna, Austria.

PETER WEIBEL studied medicine, literature, film, philosophy, and mathematics (modal logics) in Vienna and Paris. Besides his activities as artist and curator his publications about art and media theory earned him international renown. Since 1976 he has lectured widely at universities and academies in Europe and the US. After heading the digital arts laboratory at the Media Department of New York University in Buffalo from 1984 to 1989, he founded the Institute of New Media at the Städelschule in Frankfurt-on-Main in 1989. Between 1986 and 1995, he was in charge of the Ars Electronica in Linz as

artistic consultant and later artistic director. From 1993 to 1999, he was curator at the Neue Galerie Graz and commissioned the Austrian pavilions at the Venice Biennale. Peter Weibel has been Chairman and CEO of the ZKM Center for Art and Media Karlsruhe, since 1999.
Address: Center for Art and Media, Lorenzstr. 19, 76135 Karlsruhe, Germany. `www.zkm.de` E-mail: weibel@zkm.de

Quantum Mechanics and Non-Ordinary Statistical Mechanics

by Paolo Grigolini

Abstract

We argue that the de-coherence theory mimicking the occurrence of events in quantum mechanics does afford a satisfactory, albeit perhaps philosophically questionable, explanation of the emergence of classical mechanics, only in the case when both the system of interest and its environment fit conditions of ordinary statistical mechanics. In the case of anomalous statistical mechanics the de-coherence theory meets problems, and cannot maintain the promise of properly ensuring the transition to the classical regime. We examine the experimental case of intermittent fluorescence, with a non-Poisson distribution of waiting times in the "on" and "off" states, and we show that this anomalous condition becomes incompatible with the ordinary prescriptions of quantum statistical mechanics. We show that a possible way to address these issues rests on leaving the density perspective and on adopting the prescriptions of continuous random walk. We set for the advocates of de-coherence theory the challenge of reproducing these results from within a rigorously quantum mechanical approach, with no events, and use of only unitary transformation.

1 Introduction

In the last few years there has been an increasing interest in settling the paradoxes emerging from the attempts at interpreting the mathematical objects of quantum mechanics as elements of reality. Ord [1] has clearly accounted for the non-intuitive aspects of the double-slit experiment extending classical mechanics to allow fractal trajectories. Nottale [2] has developed the theory of scale relativity as an attempt at developing the consequences of releasing the usual assumption of space-time differentiability. El Naschie [3] made attempts at affording a unified view of physics and cosmology using the complex time and fractal space perspective. El Naschie, who is a champion of a new physics where the second principle is an incontrovertible fact of Nature, rather than an illusion, showed [4] that the Cantorian space can serve as a geometrical model for a space-time support of the thermodynamical approach.

This paper aims at addressing the problem from a different perspective. Rather than proposing a new approach to interpret with realistic models the facts of nature that quantum mechanics accounts for so well, we focus our attention on some new experimental effects that might require an extension of ordinary quantum mechanics. It is well known that the founding fathers of quantum mechanics introduced the concept of wave-function collapse to establish a connection between quantum mechanics and reality. The big issue regards the wave-function collapse, and the associated phenomenon of quantum jumps. Does the wave-function collapse have to be established on the basis of a new physics, as the work of Refs. [1, 2, 3, 4] seems to suggest, or it can be accounted for without going beyond the ordinary quantum mechanical theoretical framework? Many physicists believe that decoherence theory explains wave-function collapses without changing quantum mechanics, and implicitly, we think, without adopting the new fractal perspectives [1, 2, 3, 4]. Decoherence theory was born in 1970 with the seminal work of Zeh [5]. The work of Zurek [6] gave a decisive contribution to make it very popular and so widely accepted as to lead the authors of Ref. [7] to conclude that the decoherence theory renders obsolete the hypothesis of wave-function collapse made by the founding fathers of quantum mechanics. We want to show that this conclusion would be generally correct, if ordinary statistical physics were an exhaustive representation of reality. It is not so. Complexity is becoming a new important paradigm of science, and complexity can be defined as a condition of anomalous statistical physics [8]. The complexity vision, on the other hand, is rendered necessary by the fact that there are many processes, in physics, as well as in biology, sociology, physiology, and so on, which go beyond the limits of ordinary statistical mechanics. Thus, we think it to be worth to discuss whether or not the successful achievements of decoherence theory can be extended from ordinary to anomalous statistical mechanics. We plan to show that special physical conditions emerge from anomalous statistical physics that might not be properly described by quantum mechanics, in spite of the ingenuity of the advocates of decoherence theory.

The outline of this paper is as follows. In section 2 we show that decoherence theory rests on the assumption that the environment of a system of interest obeys ordinary statistical physics. If the environment yields an anomalous process of diffusion, the condition of fast de-coherence can be violated. In Section 3 we review some earlier work in the field of quantum chaos, which affords a second example of processes departing from ordinary statistical mechanics, and consequently creating some problems to the decoherence perspective. In this second example the condition of ordinary statistical physics is violated by the system. We also make an excursion into the field of quantum jumps to express the conjecture that the advocates of de-coherence theory would interpret the failure of de-coherence theory in this case as the macroscopic manifestation of quantum coherence. Section 4 addresses the intriguing problem of the statistical analysis of a symbolic sequence generated by the balance of a coherence generating in-phasing process and a frequent measurement producing wave-function collapses. In the Poisson case the result of this statistical analysis is compatible with a quantum master equation, thereby confirming the validity of de-coherence theory, in this case. To find a possible conflict between the trajectory picture and de-coherence theory we move to considering the non-Poisson case. This is done in Section 5, which illustrates the formal connection between Continuous Time Random Walk (CTRW) and Generalized Master Equation (GME). The GME is a structure compatible with a quantum mechanical derivation. In Section 6 we show that if complexity is derived from the modulation of a Poisson process, as assumed by Beck [9], then the de-coherence theory can be made compatible with complexity. In Section 7 we set the challenge for the advocates of de-coherence theory of taking into account the case where the fluorescent intermittent process obeys renewal theory.

2 Decoherence Theory

As discussed in Ref. [10], decoherence theory implies that the quantum superposition of two macroscopically distinct states, $|A>$ and $|B>$, is forbidden by the entanglement process

$$(|A>+|B>)|E> \rightarrow |A>|E>_A + |B>|E>_B, \qquad (1)$$

where $|E>$ is the initial environment state and $|E>_A$ and $|E>_B$ are the states expressing how the environment adapts itself to the system's state $|A>$ and $|B>$, respectively. As strange as this property might seem, that a big system adapts itself to a much smaller, it is a genuine quantum mechanical property that occurs with no significant exchange of energy, if the energy of $|A>$ is identical to that of $|B>$, so ensuring the condition for the second principle of thermodynamics to apply. This is fine and takes place naturally as an effect of a unitary transformation. It is interesting to examine this property from the perspective of the reduced density matrix. The reduced density matrix ρ corresponding to the left hand side of 1 is that of a pure state, fitting the condition $\rho^2 = \rho$, thereby leading to a vanishing von Neumann's entropy. When the entanglement process, indicated by the arrow of 1, is completed, the reduced density matrix becomes mixed, thereby producing a finite von Neumann entropy. Thus, if we look at this entanglement process from within the perspective of the reduced density matrix, we see the density matrix becoming diagonal with respect to the basis set of the states $|A>$ and $|B>$. These two states are the eigenstates of an observable, for instance position, corresponding to two distinct eigenvalues of this observable, and are consequently orthogonal. We also find that the von-Neumann entropy of the reduced density matrix reaches its maximum value. For this important condition to apply, we need the following condition

$$< E_A | E_B > = 0. \quad (2)$$

In a recent article [11], Adler questioned the validity of this picture, on the basis of the fact that in his opinion Eq. (2) conflicts with the claim of unitary transformation made by the advocates of decoherence theory. Actually, the work of Bonci et al [12] seems to confirm the validity of Eq. (2). These authors studied the special case where the environment of the the two-state systems, with eigenstates $|A>$ and $|B>$, is an oscillator. The two states $|A>$ and $|B>$ are the two states of a 1/2-spin system, with an in-phasing coupling of intensity ω_0, which forces the system to establish a linear superposition of these two states.

It is enough to assign to the quantum oscillator a coherent physical condition corresponding to exciting infinitely many oscillator eigenstates so as to make the reduced density matrix to become diagonal, with a corresponding entropy increase. The process implies that after reaching the condition of maximum entanglement, corresponding to Eq. (2) the system return to the initial condition. However, if we go closer and closer to the condition $\omega_0 = 0$, the recursion times become infinite, and the condition of Eq. (2) becomes irreversible. Actually, the time evolution of the states $|E_A>$ and $|E_B>$ is not independent of the system of interest, and this is probably the reason why the remarks of Adler on the unitary transformation do not apply. It is worth showing explicitly how to establish the connection between de-coherence theory and ordinary diffusion processes, following the approach of Ref. [10]. Let us assume that the Hamiltonian under study is

$$H = E(|A><A| + |B><B|) + g((|A><A| - |B><B|)\xi + H_E. \quad (3)$$

The time evolution of the variable ξ depends on the Hamiltonian H_E, referring to the environment, whose explicit expression is not important for our discussion. We limit ourselves to imagining that the environment consists of virtually infinitely many degrees of freedom. The two states $|A>$ and $|B>$ are degenerate. The results of our discussion are independent of the value assigned to E, which, for simplicity, is assumed to vanish, $E = 0$. It is convenient to use the interaction picture to represent the time evolution of the wave-function $|\psi(t)>$. By writing the formal solution of the Schrödinger equation in the interaction picture, and returning to the laboratory reference frame we obtain

$$|\psi(t)> = exp(-\frac{iH_B t}{\hbar})exp[-ig(|A><A| - |B><B|)\int_0^t dt' \xi(t')]|\psi(t)>, \quad (4)$$

where
$$\xi(t) \equiv exp(iH_E t)\xi exp(-iH_E t). \tag{5}$$

Let as assume that the initial condition is
$$|\psi(0)> = \frac{1}{\sqrt{2}}(|A> + |B>)|E>, \tag{6}$$

where the vector $|E>$ denotes the initial environment state. At the initial time system and environment are not entangled, and are statistically independent. We obtain immediately for the time evolution of Eq. (4) the following form

$$|\psi(t)> = |A>|E(t)>_A + |B>|E(t)>_B, \tag{7}$$

where
$$|E(t)>_A = exp(-\frac{iH_B t}{\hbar})exp[-i\frac{g}{\hbar}\int_0^t dt'\xi(t')]|E> \tag{8}$$

and
$$|E(t)>_B = exp(-\frac{iH_B t}{\hbar})exp[i\frac{g}{\hbar}\int_0^t dt'\xi(t')]|E> \tag{9}$$

According to de-coherence theory the human observer perceives only the system of interest. Thus, we have to evaluate first the total density matrix corresponding to the wave-function of Eq.(4) and derive the contracted density matrix out of it, $\rho(t)$. Thus, we get

$$\rho(t) = |A><A| + |B><B| + |A><B|\Phi_{AB}(t) + |B><A|\Phi^*_{AB}(t), \tag{10}$$

where
$$\Phi_{AB}(t) = <E|exp[-2i\frac{g}{\hbar}\int_0^t dt'\xi(t')]|E>. \tag{11}$$

The function $\Phi_{AB}(t)$ should be evaluated with a rigorous quantum mechanical calculation. In an earlier work [10] we noted that de-coherence theory simplifies this task by assuming that the time dependent quantum operator $\xi(t)$ is a stochastic classical variable, so that the function $\Phi_{AB}(t)$ becomes a characteristic function. By assuming that $\xi(t)$ is an ordinary Gaussian and uncorrelated fluctuation, it is shown that

$$\Phi_{AB}(t) = exp(-2Dt), \tag{12}$$

where
$$D = g^2 <\xi^2>/\hbar^2. \tag{13}$$

In the specific case where $|A>$ and $|B>$ are two coherent states of the same macroscopic oscillator, located at a distance $<\Delta x>$ apart from one another, the prescription of the Hamiltonian of Eq. (3) can be maintained provided that the coupling g is expressed in terms of the distance Δx. This yields for the de-coherence time the expression

$$t_D = \frac{\hbar^2}{2\Sigma(\Delta x)^2}, \qquad (14)$$

where Σ is proportional to the product of temperature and friction of the macroscopic oscillator, as shown in Ref. [10]. The important fact that for a macroscopic distance and for a macroscopic oscillator the de-coherence time becomes so small as to ensure that there is no room for the superposition condition $|A> +|B>$ in classical physics.

One might question the claims of decoherence theory on the basis of the fact that the reduced density matrix becoming diagonal, if an individual system perspective is adopted, might imply blurring [13] rather than a wave-function collapse. The authors of Ref. [14, 15] made the conjecture that a real wave-function collapse might occur, as a result of the enhancement of spontaneous collapses, meant as true correction to ordinary quantum mechanics. An attractive example of theory of this kind is given by the work of Ghirardi, Pearle and Rimini [16], with a wise choice of space and time scale of the process of spontaneous collapse as to produce quite negligible corrections to ordinary quantum mechanics if the elementary microscopic constituents are individually studied. The macroscopic wave-function collapse is the result of an enhancement, due to the ordinary interactions, the ultimate effect of which is to yield a genuine wave-function collapse, with no significant statistical effects. However, these arguments do not represent a real failure of decoherence. Rather, from a practical point of view, this would be a success of decoherence theory, and the practical role of spontaneous collapses might not properly appreciated, even because it would rest on complicated calculations leading to results identical to those predicted by decoherence theory with a much easier algebra. This is a consequence of the fact that the recourse to ordinary statistical mechanics makes it impossible to defeat decoherence theory.

We would like to attract the attention of the reader to the case when the environment is a source of anomalous diffusion. Paz, Habib and Zurek [17] studied the de-coherence process generated by a supra-ohmic bath, but they did not find any problem with the adoption of the de-coherence theory. It is convenient to devote some attention to the case when the fluctuation ξ is a source of Lévy diffusion [18]. If the fluctuation ξ is an uncorrelated Lévy process, the characteristic function decays exponentially again, and the only significant change is that the parameter Σ appearing in Eq. (14) would have a sub-linear dependence on temperature. However, it is more interesting to consider the case when ξ yields a Lévy walk rather than a Lévy flight [19]. The Lévy walk is a paradigmatic example of complexity fitting the definition proposed in Ref. [8], where complexity is meant to be a condition of transition from dynamics to thermodynamics, lasting for an infinite time. In fact, the Lévy walk has the following property: The diffusion process departs from the condition of mono-scaling predicted by the generalized central limit theorem, and lives forever in a state intermediate between multi-scaling (dynamics) and mono-scaling thermodynamics. In the multi-scaling case, the probability density of the diffusing variable keeps changing its form, thereby making it impossible for us to define a condition equivalent to thermodynamic equilibrium. In this case the characteristic function $\Phi_{AB}(t)$, after a fast exponential phase, moves to an oscillatory regime, with a slowly decreasing intensity, which, in practice, takes an infinite time to vanish.

As we shall see in Section 7, this is a condition that might create some problems to the de-coherence theory and should deserve some attention. It is important to point out that some more research work is necessary to turn this property into a compelling evidence of the de-coherence theory crisis created by the anomalous statistical condition. However, it is convenient to devote here some more room to illustrating what do we mean by anomalous, or complexity, condition. First of all, we remind the reader that scaling, either normal or anomalous, is defined by

$$p(x,t) = \frac{1}{t^\delta} F(\frac{x}{t^\delta}), \qquad (15)$$

where $p(x,t)$ is the probability distribution density of the diffusion process. Anomalous statistics implies either $\delta \neq 1/2$ or $F(y)$ being a non-Gaussian function of y, or both conditions. This property allows us to define a form of equilibrium for a diffusion process, whose actual distribution keeps broadening forever, in an apparent conflict with equilibrium. The function $F(y)$ is the expression of the thermodynamic equilibrium to which we have been referring to. When this condition is realized, regardless of the shape of $F(y)$ and the value of δ, the process of de-coherence is very fast. However, in the case of Lévy walk, the work of Ref. [19] proves that this condition is reached after a transient infinitely extended in time. Throughout the whole complex condition, the de-coherence process is extremely slow, and might leave room for the emergence of macroscopic quantum effects. Eventually, when the scaling condition is reached, the de-coherence process becomes very fast again. In conclusion, the condition of complexity fitting the definition of Ref. [8] seems to prevent the de-coherence process from producing a fast annihilation of quantum coherence.

3 Transition from the quantum to the classical dominion

Quantum chaos is a field of research whose aim is to study the dynamics of those nonlinear systems that would be characterized by deterministic chaos in the classical limit. The interested reader can consult excellent review papers, such as Ref. [20], where the definition of quantum chaos is given through the spectral properties, thereby applying also to systems with no classical counterparts. However, here we limit ourselves to discussing a quantum process, with a classical limit: the kicked rotor [21]. This is a rotor kicked at regular intervals of time by an impulsive torque with a strength proportional to $K\sin\theta$, where θ is an angle denoting the rotor orientation. At any impulsive kick the rotation momentum x changes by a a given quantity ξ. Thus after many kicks the system dynamics become indistinguishable from

$$\frac{dx}{dt} = \xi(t). \tag{16}$$

In fact, if the control parameter K is large enough, the variable ξ is an uncorrelated fluctuation, and, in the long-time limit, can be thought of as a noisy function of the continuous time t. The solution of Eq. (16) yields results agreeing with ordinary statistical mechanics, namely, a diffusion process making the second moment of x increase as a linear function of time. However, in the quantum case this linear increase has an upper bound in time. At times of the order of

$$t_L \propto \frac{K^2}{\hbar^2}, \tag{17}$$

the second moment of x stops increasing. In the case of a macroscopic rotor the momentum localization might be considered a macroscopic manifestation of quantum mechanics, consequently generating another quantum mechanical mystery. In fact, a macroscopic kicked rotor is expected to obey classical physics, and so its behavior should be identical to that of a classical kicked rotor, without localization. An obvious way to settle this paradox is by observing that for a really macroscopic kicked rotor the action to $\hbar$ ratio is so large, as to make the localization time of Eq. (17) infinitely much larger than the time duration of any realistic experiment. This explanation is not quite satisfactory because the localization process is the consequence of latent quantum memory that survives for a virtually infinite time, even if it does not yield physical effects at times shorter than t_L. The decoherence theory affords a more attractive explanation [22]. Adopting the Wigner formalism it is possible to express the quantum mechanical problems in terms of the classical phase space, and the Wigner quasi-probability is expected to remain definite positive till to the moment of the quantum transition occurring, according to the estimate of Ref. [22], at the time

$$t_\chi = \frac{1}{\lambda} ln(\chi/\hbar), \qquad (18)$$

where χ is a scale parameter proportional to the non-linear interaction. However, Zurek and Paz show [6, 17] that the environmental fluctuations, as weak as they are, can erase the quantum mechanical coherence, thereby preventing the birth of quantum interference effects that should take place at the time scale of Eq. (18). In the specific case of the kicked rotor the time necessary to kill the quantum mechanical coherence is given by [23]

$$t_D = \frac{\hbar^2}{2\sigma(\Delta x)^2}, \qquad (19)$$

where $\Delta x = K^2/\hbar$ and σ is the intensity of environmental fluctuation. It is possible to prove [22] that in realistic cases $t_D < t_\chi$, thereby preventing the localization process from occurring. This interpretation is attractive: at times $t > t_D$ there is no more latent quantum memory, and the system becomes genuinely classical. The decoherence theory exorcizes the emergence of macroscopic quantum effects, regardless of how far this goes into the future. If $\hbar$ is much smaller than the classical action, the system is classical, and no localization is permitted. Only ordinary diffusion is possible, as a consequence of the fact that in the classical limit, as we have seen, the variable ξ of Eq. (16) is an uncorrelated noise, with no memory whatsoever.

As pointed out in Section 1, we want to explore the case of anomalous statistical mechanics. Anomalous diffusion is generated by Eq. (16) when ξ departs from the condition of uncorrelated noise. In the case of the kicked rotor this anomalous condition is realized by assigning to the control parameter K special values. These special values make two accelerator islands appear in the phase space of the kicked rotor [24], embedded in a chaotic sea. The surface of separation between the accelerator island and the chaotic sea is sticky. This means that the trajectory undergoes an erratic motion in the chaotic sea, and, then, from time to time, as an effect of erratic diffusion, sticks to the surface of one of these two islands. Throughout the extended time of sojourn at the surface with one of these two islands, the momentum keeps increasing by the same quantity, either W or $-W$, depending on the island. The important fact is that the waiting time distribution in one of the two states, $\psi(t)$, has the following time asymptotic property

$$lim_{t\to\infty}\psi(t) \propto \frac{1}{t^\mu}, \qquad (20)$$

where $2 < \mu < 3$. The condition of ordinary statistical mechanics, the Poisson condition, corresponds to an exponential waiting time distribution, namely, to $\mu = \infty$. Thus, we are in the presence of a physical condition where anomalous rather than normal statistical mechanics is involved.

In the case studied in Ref [25], $K = 6.9115$ and $\mu = 2.667$. The reasons why this separation surface yields the anomalous waiting distribution are well known, and the reader can find a detailed discussion in Ref. [24], for instance. The separation surface actually is not a curve with vanishing width. It is rather a layer of finite size, with channels that make it possible for the surrounding sea to penetrate the layer. The surrounding sea generates these channels through ramification that generates the branch channels of decreasing size as moving towards the interior of the layer. As a consequence of this geometric structure, in the classical case a trajectory sticks to the layer, and thus to the border with one of the two accelerator islands, for long times, with a distribution density given by Eq. (20). In the laboratory frame the second moment of x is proportional to $t^{4-\mu} = t^{1.333}$, which is significantly faster than the linear in time increase, reflecting the prescription of ordinary statistical mechanics. The authors of Ref. [25] confirmed the new effect discovered by the authors of an earlier paper [26]. This is as follows. The second moment of the momentum of quantum kicked rotor grows faster than in the ordinary case for a while, then, at a given time it makes a transition to a condition of linear in time growth. This effect can be accounted for as follows. Let us denote by $\Xi(0)$ the area of the portion of the separation surface, directly arrived by trajectories starting from the chaotic sea. In accordance with the assumption that the kicked rotor is virtually classical, we assume that $\hbar$ is much smaller than $\Xi(0)$. Notice that, according to the theory accounting for the origin of the inverse power law of Eq. (20), the trajectory penetrates within the boundary regions, through channels of decreasing size. Thus, the area $\Xi(t)$ is a decreasing function of time. On the other hand, quantum mechanics turns the bunch of classical trajectories into a wave function with the quantum uncertainty, $U(t)$, increasing as $exp(\lambda t)$. The correspondence with classical physics is lost when

$$\Xi(t) = U(t). \tag{21}$$

The reason why the condition of Eq. (21) yields the breakdown of the correspondence between quantum and classical mechanics is evident. The quantum wave function can be identified with a trajectory if it is enough sharp, namely if $U(t) < \Xi(t)$. When the width of the wave function becomes as large as the width of the channel, within which the wave function moves, the wave function dynamics begins depending on the structure of the surrounding phase space, and the correspondence is broken.

Using a geometrical model making $\Xi(t)$ decrease exponentially with time [26], the condition of Eq. (21) is shown to occur at time $t = t_B$, where

$$t_B = \frac{1}{\lambda} ln(1/\hbar). \qquad (22)$$

This prediction has been confirmed by the results of Refs. [27] and [28]. In fact, the numerical result of Ref. [27] indicates that the waiting time distribution of Eq.(20) has an exponential truncation, this being an effect of the tunneling from the border between chaotic sea and accelerator island, back to the chaotic sea. The authors of Ref. [26, 25] argue that the quantum induced recovery of ordinary diffusion is followed by a corresponding localization process.

How to use the decoherence theory to annihilate these quantum effects? There are problems. In fact, in classical physics the adoption of environmental noise produces a departure from anomalous diffusion [29, 30]. Thus, the assumption of the decoherence theory that there are no isolated systems, and that we have always to consider the influence of environmental fluctuations, would kill anomalous diffusion. Furthermore, the numerical results of Ref. [25] show that the quantum-induced transition from anomalous to ordinary diffusion is a quantum effect more robust than the localization phenomenon itself. This indicates that in this case the presence of a weak environmental fluctuation is not enough to re-establish the correspondence principle.

We are now in the right position to reach a preliminary conclusion. Although the decoherence theory is an attractive and efficient way of defeating the emergence of quantum effects at a macroscopic level, the authors of Ref. [13] did not feel comfortable with it. The reason is that when the observer has the impression that a wave-function collapse occurs, actually the quantum mechanical coherence is becoming even more extended and macroscopic, since it spreads from the system to the environment, Eq. (1). It can be shown [31] that the measurement process itself has the effect of making microscopic coherence grow till to extend to the whole universe. In the measurement process $|A>$ and $|B>$ are two eigenstates, with eigenvalues α and β, respectively, of a given observable O pertaining to a microscopic system a. The environment of the microscopic system a is the experimental apparatus itself, with a pointer up, $|E>_A$, and down, $|E>_B$, indicating to the macroscopic observer that the observable to measure has the value α and β, respectively. If the microscopic system is in a linear superposition of $|A>$ and $|B>$, the experimental apparatus would end up in the corresponding linear superposition of $|E_A>$ and $|E_B$, a fact that would correspond to the emergence of quantum mechanics at the macroscopic level. Decoherence theory has to be invoked at this level, by introducing the environment S of the measurement apparatus, with two orthogonal states $|S>_A$ and $|\bar S>_B$. Thus, we have to assume that the observer cannot see the linear superposition of s, because this is a microscopic system. The observer cannot see the linear superposition of S, either, because the observation focuses only the experimental apparatus. This latter assumption is subtly related to considering the second principle, and the transition from quantum to classical physics as well, as a consequence of human limitation rather than as an objective fact of nature.

For these reasons, the authors of Ref. [13] argue that de-coherence theory does not produce genuine wave-function collapses, but rather an effect determined by the limited information available to the observer, a property that they define *subjective collapse*. They notice [13] that the theory of Ref. [16] generates genuine wave-function collapses. This theory is a generalization or extension of quantum mechanics, turning the wave-function collapse assumption of the founding fathers of quantum mechanics into an essential dynamical ingredient. There should be no concern for losing the wonderful achievements of quantum mechanics, for more than 100 years [7]. In fact, thanks to a wise choice of two new constants [16], the new Schrödinger equation, with a stochastic correction, is almost always equivalent to the ordinary Schrödinger equation. The corrections to quantum mechanics are activated by the creation of the cooperative effects triggered by the measurement process [13]. As pointed out in Refs. [14, 15], the physical processes necessary to produce de-coherence might have the effect of activating genuine wave-function collapses. Thus, the main problem here is not so much the risk of losing the benefits of quantum mechanics, but rather it is how to make the theory of objective wave-function collapses distinguishable from de-coherence theory. In fact, from a statistical point of view, the picture adopted by the authors of Ref. [13] is essentially equivalent to the de-coherence theory.

On the same token the increasing interest for the stochastic Schrödinger's equation to study many-body problems [32, 33, 34] does not imply a conversion of these authors to the theory of genuine wave-function collapses. Some of these authors [34] did succeed in realizing the important condition of norm conservation, which would be an appealing property from within the perspective of wave function collapses (for a discussion of the wave-function collapse with no norm-conservation, see for instance, Ref. [35]). However, the main aim of these authors is to create an efficient computational tool with a stochastic picture that yields results that are statistically equivalent to the Lindblad master equation [37], a well known Markov master equation. The norm preserving stochastic Schrödinger equation, statistically equivalent to the Lindbland master equation, was pioneered by Gisin and Percival [36] and is formally equivalent to the stochastic Schrödinger equation of [16], a fact reinforcing our conviction that it is very difficult, if not impossible, to distinguish subjective from objective wave-function collapses.

In conclusion, the condition of ordinary statistical physics makes the decoherence theory a valuable perspective, and an attractive way of deriving classical from quantum physics. The argument that the Markov approximation itself is subtly related to introducing ingredients that are foreign to quantum mechanics [38] cannot convince the advocates of decoherence theory of the needs of looking for a new physics. The only possible way of converting a philosophical debate into a scientific issue, as suggested by the results that we have concisely reviewed in this section, is to study the conditions of anomalous statistical mechanics. In the next sections we shall explore with more attention these conditions.

What about the quantum breakdown of anomalous diffusion [26, 25]? It seems that later theoretical work [28] confirmed the results of Ref. [26, 25]. The discovery that the new effect is robust against environmental fluctuations [25], if taken into account by the researchers in this field, might have the effect of triggering some experimental research work in this direction. However, it is expected that the experimental observation of this effect would be thought of as the discovery of an interesting way to produce macroscopic quantum effects: a strange condition that would not be perceived as a failure of quantum mechanics. In other words, de-coherence theory would leave open some special channels for the emergence of macroscopic quantum effects. In the remainder part of this paper we shall explore other physical processes that might yield instead the crisis of quantum mechanics itself as well as of de-coherence theory.

4 Trajectory and Density Entropies

The so called Kolmogorov-Sinai (KS) entropy [39] is a property of a time sequence of symbols and can be interpreted as the mean entropy increase per unit of time. In the case of a dynamic system the sequence of symbols is generated by a trajectory running through a phase space divided into many cells of finite size and labeled with given symbols. In this case this form of entropy can be related to the Lyapunov coefficient [40]. If the density approach is used, we make the conjecture that the spreading of the density distribution is proportional to the Lyapunov coefficient. Thus, the ordinary Gibbs entropy is expected to increase linearly in time, with a rate that turns out to be proportional to the KS entropy. This vision, advocated by Latora and Baranger [41], has attracted the attention of many researchers. We note that the perspective advocated by these authors is based on the implicit assumption that trajectories are more fundamental than densities.

This is a crucial aspect subtly related to the main goal of this paper. For this reason, it is worth of an extended discussion. To a first sight, questioning the equivalence between trajectories and densities does not make sense, given the fact that the density equations (Liouville, Liouville-like and Frobenius-Perron equations alike) are built up for the specific purpose of reproducing the time evolution of a bunch of trajectories. In literature the existence of a possible conflict between densities and trajectories can be found in the work of Petrosky and Prigogine [42]. Although disconcerting, their claim makes sense. In fact, according to these authors the adoption of densities, and consequently of the Liouville approach, implies the diagonalization of matrices. These authors pointed out that there exists a sort of equivalence between Hamiltonian systems with infinitely many degrees of freedom and low-dimensional chaotic systems. In both cases the dynamic operator driving the time evolution of probability density, expanded on a suitable basis set, becomes a matrix of infinite size. In both cases recourse to the method of analytical continuation is done, this being the source of irreversibility within the context of physical laws which are invariant by time reversal. Thus they establish a distinction between trajectories and densities and consider the latter more fundamental than the former. In this paper, as we shall see in detail in Section 7, we reverse this perspective, and we prove that the existence of critical events generates the important aging property that is forcing us to revise the approaches currently adopted to study the interaction between a system and an external perturbation. However, the view of Petrosky and Prigogine at this stage serves the very useful purpose of making more convincing our arguments about the importance of using density equation to study the time evolution of the Gibbs entropy.

There exists another deep reason in favor of studying the Gibbs entropy, or its quantum mechanical counterpart, the von Neumann entropy [43], by means of the corresponding density equation. In quantum mechanics there are no trajectories. In quantum statistical mechanics we use the quantum Liouville equation, which drives the time evolution of the statistical density matrix. The advocates of decoherence theory interpret a contracted quantum Liouville equation in terms of stochastic trajectories. As pointed out in Section 3, any claim about the individual system observation is not an argument convincing enough for the advocates of decoherence to abandon their view. This is so because, even if the observation of quantum jumps [44] implies the occurrence of events, and can be judged to be equivalent to the experimental observation of wave-function collapses, yet the advocates of decoherence adopt a statistical perspective, which is equivalent to making averages on the Gibbs ensemble, even if the individual systems of this ensemble are directly observable. The only possible way out of this ambiguous condition seems to be offered by the breakdown of the equivalence between the trajectory and density picture [42]. It is worth to explore this issue, even if this discussion will lead us to the opposite conclusion of considering trajectories more fundamental than densities. To shed light into this intriguing issue, we are convinced that if a density perspective is adopted, then use of the corresponding equation of motion, and only of that, has to be made. As we shall see, the adoption of this perspective will lead us to focus our attention on the systems characterized by anomalous rather than ordinary statistical mechanics.

The first reason that led Latora and Baranger to evaluate the time evolution of the Gibbs entropy by means of a bunch of trajectories moving in a phase space divided into many small cells is the following: In the Hamiltonian case the density equation must obey the Liouville theorem, namely it is a unitary transformation, which maintains the Gibbs entropy constant. However, this difficulty can be bypassed without abandoning the density picture. In line with the advocates of decoherence theory, we modify the density equation in such a way as to mimic the influence of external, extremely weak, fluctuations [45]. It has to be pointed out that from this point of view, there is no essential difference with the case where these fluctuations correspond to a modified form of quantum mechanics [16].

There exists a second reason why Latora and Baranger have been forced to depart from the adoption of a density equation, thereby rather adopting the supposedly equivalent time evolution of a bunch of trajectories. This is due to the fact that the Lyapunov coefficients are local and might change with moving from one point of the phase space to another. It is important to stress this second reason because it is closely related to the directions to follow to reveal by means of experiments the breakdown of the density perspective, and with it of quantum mechanics, in spite of the fact that so far the predictions of quantum mechanics have been found to fit very satisfactory the experimental observation.

This form of disagreement between densities and trajectories has been discussed in detail by Bologna $et\ al$[45]. They studied the Manneville map [46],

$$x_{n+1} = x_n + x_n^z (mod 1), \qquad (23)$$

with $z > 1$. This important map, when $z = 1$ becomes identical to the Bernoulli map [47], which is characterized by the same Lyapunov coefficient,

$ln2$, throughout the whole definition interval $I = [0,1]$. For any $z > 1$ the interval splits into two parts, one containing $x = 0$, laminar region, and one containing $x = 1$, chaotic region. The chaotic region is characterized by very large Lyapunov coefficients, while the laminar region is filled with Lyapunov coefficients becoming increasing smaller as x comes closer to $x = 0$. In this condition, it is impossible to make the rate of the Gibbs entropy increase become identical to the KS entropy. Notice that when $z > 2$ the Manneville map does not admit an invariant measure anymore, and with it does not admit any ergodic property, either. However, at $z < 2$, where the KS entropy h_{KS} is known to be finite ($h_{FS} > 0$) [48], and the system is ergodic, we are in the presence of this interesting scenario. A single trajectory explore in time the whole interval I, chaotic and laminar region alike. This corresponds to a condition of thermodynamic equilibrium, which, in the density framework is given by the equilibrium distribution. If we adopt the density prescription, we must establish at the initial time an out of equilibrium condition, namely a delta of Dirac distribution located in some point of the interval I corresponding to a well defined Lyapunov coefficient. However, as the distribution starts spreading, new regions, with different Lyapunov coefficient are involved, thereby making it impossible for the initial growth of the Gibbs entropy to increase with a rate corresponding to a single Lyapunov coefficient.

This condition plays such an important role for our discussion that it is worth to deserve some room to the discussion of an idealized version of the Manneville map, where the laminar region is characterized by vanishing Lyapunov coefficients and the chaotic region, in a sense, by an infinite Lyapunov coefficient [45, 49]. In fact, for calculation and conceptual purposes it is convenient to replace the Manneville map with the very simple model given by

$$\frac{dy}{dt} = \lambda y^z, \qquad (24)$$

with the variable y moving in the interval $I \equiv [0,1]$. The variable moves from the left to the right and when it reaches the point $y = 1$, it is injected back with uniform probability in the interior of the interval I. It is straightforward to prove that the waiting time distribution is given by

$$\psi(t) = (\mu - 1)\frac{T^{\mu-1}}{(t+T)^{\mu-1}}, \qquad (25)$$

where

$$\mu = \frac{z}{(z-1)}. \qquad (26)$$

In the case $\mu > 2$ the waiting time distribution is characterized by a finite first moment, namely by the mean time

$$<t> = \frac{T}{\mu - 1}.\qquad(27)$$

This means that an invariant distribution exists, and that after a transient, of infinitely large duration if $\mu < 3$, the rate of random events is constant. In the case of the Manneville map, the mean Lyapunov coefficient is finite. In this case the Kolmogorov-Sinai entropy, namely the mean Lyapunov coefficient, is the sum of two sets of local Lyapunov cofficients. The Lyapunov coefficients of the former set have very small values, corresponding to the deterministic laminar region, of the idealized model, with vanishing Lyapunov coefficient, and the Lyapunov coefficients of the latter set have very large values, corresponding, so to speak, to the infinite Lyapunov coefficient of the idealized model. We shall use the idealized model to discuss the consequences on the quantum mechanical perspective of non-Poisson statistics.

Here, we limit ourselves to remarking that the discovery made by the authors of Ref. [45] was found to have a quantum mechanical counterpart [43]. These authors have [43] studied the case of two degenerate quantum states, $|1>$ and $|2>$, with the very simple Hamiltonian

$$H = V(|1><2| + |2><1|).\qquad(28)$$

This Hamiltonian corresponds to an inphasing process, in the sense that if we start with either the state $|1>$ or the state $|2>$, this Hamiltonian produces a coherent superposition of the state $|1>$ and $|2>$. For instance, the solution corresponding to the initial condition $|\psi(0)>= |1>$, reads

$$|\psi(t)> = cos(\frac{Vt}{\hbar})|1> - i sin(\frac{Vt}{\hbar})|2>.\qquad(29)$$

If we make the assumption that a very frequent measurement process is carried out, with a frequency $1/\tau$ much higher than the coherent oscillation frequency $\frac{V}{\hbar}$ the resulting effect is that of producing a condition of large persistence either in the state $|1>$ or in the state $|2>$. In fact, if we move from the condition $|\psi(0)>= |1>$, and almost immediately afterward we make a measurement, the probability of producing a collapse of the wave function into $|1>$ is much higher than that of a collapse into the state $|2>$. Thus, the system will sojourn for an extended time in $|1>$. However, sooner or later, a collapse into the state $|2>$ occurs. At that stage, the system begins sojourning in the state $|2>$ for an extended time. It is easy to show that the waiting time distribution is exponential. It is easy to evaluate the KS entropy and it is remarkable that, as in the case studied by Latora and Baranger [41], the von Neumann entropy increase with a rate coinciding with the KS entropy.

We notice that the same result can be found, by using the Hamiltonian

$$H_T = V(|1><2| + |2><1|) + (|1><1| - |2><2|)y + H_B, \qquad (30)$$

where y is a bath operator, and H_B is the Hamiltonian of this bath. In the special case where the bath time scale is much shorter than the coherent oscillation frequency, the method projection proposed many years ago by Zwanzig [50, 51, 52] generates a master equation that is equivalent to the wave-function collapse picture producing a connection between rate of increase of the von Neumann entropy and KS entropy.

In this condition, it is shown that a symbolic sequence is generated such as $+++++-------++....$ with $+$ and $-$ denoting the system being in the state $|1>$ and $|2>$, respectively. The waiting time distribution was found to have an exponential form, thereby implying a well defined KS entropy. The authors of Ref. [48] proved that the rate of von Neumann entropy increase, before the saturation condition corresponding to the attainment of equilibrium, is identical to the KS entropy of this symbolic sequence. This is a remarkable result confirming the statistical equivalence between the master equation and the symbolic sequence. Since the symbolic sequence can be thought of as being generated by wave-function collapses, this results support the claims that de-coherence theory makes obsolete the assumption of wave-function collapses. Thus, we find again that ordinary statistical mechanics makes de-coherence theory work properly. In Section 5 we shall explore the case of a non-Poisson symbolic sequence, to establish if this will cause or not problems to de-coherence theory.

5 Continuous Time Random Walk and Generalized Master Equation

Let us imagine the case of a random walk in a condition where the lattice has only two states, $|1>$ and $|2>$, with $|1>$ on the right, and $|2>$ on the left. The random walker generates a time series $++++----------+++----....$ Let us assume the isotropy condition. It is convenient to describe the statistics of the sojourn times with the waiting time distribution $\psi(t)$, which must fit the normalization condition

$$\int_0^\infty dt\psi(t) = 1. \qquad (31)$$

Ordinary statistical physics yields the exponential assumption

$$\psi(t) = \frac{\gamma}{2} exp(-\frac{\gamma t}{2}). \qquad (32)$$

In fact, the exponential condition ensures that all the moments of this distribution, $<t^n>$, regardless the value of the integer number n, are finite.

An inverse power law even with a very large power index μ would imply that the integer numbers $n \geq n_c$, with

$$n_c \equiv [\mu - 1], \qquad (33)$$

and $[b]$ denoting the integer part of b, would yield divergent moments. It is interesting to notice that there exists a GME equivalent to the CTRW on this two-dimensional lattice. This GME reads

$$\frac{d}{dt}\mathbf{p}(t) = -\int_0^t \Phi(t-t')\mathbf{K}\bullet\mathbf{p}(t')dt', \qquad (34)$$

where $\mathbf{p}(t)$ is a two-dimension vector, whose components, $p_1(t)$ and $p_2(t)$, denote the probability of finding the random walker either in $|1>$ or in $|2>$.

In accordance with the earlier results of the well known paper by Kenkre, Montroll and Shlesinger [53], the connection between the CTRW and the GME establishes that [54]

$$\hat{\Phi}(u) = \frac{u\hat{\psi}(u)}{(1-\hat{\psi}(u))}. \qquad (35)$$

If we adopt the Poisson assumption of Eq. (32), Eq. (35) makes $\hat{\Phi}(u)$ independent of u, thereby turning the memory kernel $\Phi(t)$ into a delta of Dirac. This proves that the Poisson waiting time distribution of sojourn times is equivalent to the Markov master equation [55]. Since the Markov master equation is judged to be compatible with ordinary quantum mechanics, this equivalence is in line with the widely accepted view that quantum jumps do not call for any modification of quantum mechanics. We see again that this supports our claim that the wave function collapse can be realized from within the ordinary quantum mechanics, provided that we rest on ordinary statistical mechanics.

We know, however, that there exist quantum processes, the so called blinking quantum dots [56, 57, 58], where the waiting time distribution is not Poissonian, rather an inverse power law with index μ. How to handle this new condition? One might imagine that the GME can be derived from some special Hamiltonian, given the fact that the GME of Eq. (34) has a structure compatible with the one emerging from the Zwanzig projection formalism.

Is that really possible? To the best of our knowledge there does not exist yet a derivation of this kind. We do not know yet if this is possible or not. However, we are already in a position to prove that if this derivation is possible, a new perspective must be adopted anyway. In fact, it has been recently shown by Sokolov, Blumen and Klafter [59] something that in our opinion is very disconcerting. This is that the response of the GME to an external perturbation does not coincide with the response to external perturbation of the corresponding CTRW.

This is the conclusion of a rigorous calculation, and is consequently unquestionable. However, it is possible to account for this interesting effect by considering another striking consequence of non-Poisson statistics, this being aging. The CTRW, and the corresponding GME depends on $\psi(t)$. However, this is true when observation and preparation begin at the same time. Let us explain this important fact. We can imagine that the material under study is characterized by a set of random walkers, all of them being at the beginning of the laminar region, as an effect of preparation. Let us imagine that preparation is made at time $t_a < 0$ and that observation begins at time $t = 0$. In this case the distribution of first sojourn times, which is essential to define the CTRW, does not coincide with $\psi(t)$ and depends on t_a [60].

This means that once the GME coinciding with the CTRW has been built up we cannot look at it as a fundamental law of nature. If this GME were the expression of a law of nature, it would be possible to use it to study the response to external perturbations. The linear response theory is based on this fundamental assumption and its impressive success is an indirect confirmation that ordinary quantum and statistical mechanics are indeed a fair representation of law of nature. But, as proved by the authors of [59], this is not more true in the case we are discussing.

More recently, the authors of [61] to properly address the evaluation of absorption and emission spectra had to go beyond the restrictions of the conventional approaches, and to rest on the notion of trajectory (random trajectory) rather than that of density, and of the corresponding Liouville-like equation.

6 Modulation or Renewal?

On the basis of the arguments of the earlier sections, we are in a position to guess which the crucial experiment challenging ordinary quantum mechanics might be. We have seen that the experimental evidence on quantum jumps is not considered to be crucial. The reason for this conviction is that the quantum jumps of Dehmelt [44] are Poissonian. We have seen that ordinary statistics do not create problems to the decoherence theory. We have to look for experimental effects where the waiting time distribution in either the "on" or the "off" state is not Poisson. Properties of this kind have been experimentally detected in the recent past in several quantum dot systems, mainly in single CdSe nanocrystals quantum dots [56], in self-assembled quantum dots [57] and in individual ZnS overcoated CdSe quantum dots [58]. The main fact assessed by these experiments is that the distribution of sojourn times in the "on" and "off" states is not exponential. It is an inverse power law with index μ around $\mu = 1.5$.

This fascinating problem attracted the attention of several groups. The remarkably interesting paper by Jung, Barkai and Silbey [63] is based on the renewal theory. This means that after an abrupt transition from the "on" to the "off" state, or vice-versa, the system loses any memory of the earlier jumps. The paper by Brokmann et al. [64] analyzes the time sequence of "off" and "on" waiting times, and discloses the existence of aging effects, these effects being a natural consequence of the renewal assumption in the non-Poisson case [60].

Other groups adopt a point of view based on the modulation of Poisson statistics [65, 58]. This means that the waiting time distribution is an exponential function of time, whose decay rate depends on time. Thus, the resulting waiting time distribution, if the dependence on time of the decay rate is sufficiently slow, can be expressed as a sum of infinitely many exponential decays, and this sum, in turn, depends on time as an inverse power law. To be more specific, let us mention that using the proposal of Beck [9] we can express the explicit form

$$\psi(\tau) = (\mu - 1)\frac{T^{\mu-1}}{(\tau + T)}, \qquad (36)$$

as

$$\psi(\tau) = \int d\gamma p(\gamma) \gamma exp(-\gamma\tau). \qquad (37)$$

It is possible [66] to assign to the distribution $p(\gamma)$ an analytical form such that $\psi(\tau)$ of Eq. (37) gets the inverse power law form of Eq. (36). Note that renewal theory means that the number τ_i, denoting the sojourn time in either the "on" or the "off" state, with the conditions "on" and "off" determined by tossing a coin, are randomly drawn from a box of numbers with the distribution of Eq. (36). The prescription of Eq. (37) implies that we draw first a sequence $\{\gamma_i\}$ with the distribution $p(\gamma)$. Then, for any γ_i we build up a long subsequence of times, randomly drawn from the corresponding Poisson distribution. If for any number $\{\gamma_i\}$, we keep fixed the numbers of waiting times, we realize an even deeper statistical equivalence between modulation and renewal. The two sequences not only yield the same inverse power law distribution of Eq. (36): In the case $\mu > 2$, they do generate also the same analytical form for the correlation function of the fluctuation ξ, with $\xi = 1$ and $\xi = -1$, for the "on" and "off" state, respectively. However, the complete statistical equivalence between the two time series requires that the times of the latter are totally reshuffled. This means that the complex nature of a process cannot be fully determined by the waiting time distribution. In the case $\mu > 2$, the additional requirement that also the two correlation functions are identical, is not enough to ensure the complete equivalence between the two processes.

At the moment of writing this paper it is not yet quite clear if the renewal theory adopted to study the intriguing case of blinking quantum dots is the only possible perspective. In principle a non-Poisson waiting time distribution can be created via modulation of Poisson processes. We expect, however, that the modulation perspective does not yield the aging effects established by the analysis of experimental data [64]. The renewal theory, on the contrary, fits the aging effects. The modulation theory does not conflict with ordinary quantum mechanics, whereas the renewal perspectives yields effects and formal structures that might be incompatible with the ordinary treatments. If these properties were proved to be incompatible with quantum mechanics, this would lead to the conclusion that a new physics should be looked for. This new physical theory might be based on a perspective advocated by El Naschie, and other scientists as well, where randomness is not an illusion, but a fundamental property of nature.

It seems that the ordinary approach to the quantum mechanical master equation rests on the use of equilibrium correlation function. Thus the CTRW method used by the authors of Ref. [61], yielding special forms of GME [54], should be made compatible with the GME, whose memory kernel is a correlation function of a bath fluctuating variable, as discussed in Ref. [62]. The measuring apparatus, mentioned in Section 3 is a 1/2-spin, corresponding to the operator $\mathbf{\Sigma_z}$, whose two eigenstates must be identified with the states $|E_A>$ and $|E_B>$ of Section 3. Let us make this operator with a bath variable that is another 1/2-spin operator, called σ_x,

$$H_{int} = G\Sigma_z \sigma_x. \tag{38}$$

The adoption of the ordinary approach to GME, yields

$$\frac{d}{dt}\rho_\Sigma = 2(\frac{G^2}{\hbar^2}) \int_0^t dt' \Phi_\sigma(t-t')[\Sigma_z \rho_\sigma(t') \Sigma_z - \rho_\sigma(t')]. \tag{39}$$

This structure is compatible with the modulation approach to the anomalous waiting time distribution, and with the adoption of the ordinary Liouville-like approach [66]. It is shown [67] that this structure implies that the higher-order correlation functions factorize according to the prescription $< 4321 > = < 43 >< 21 >$. This factorization condition, valid in the Poisson case, is broken when the statistics of the system departs from the Poisson condition [68]. In the Poisson case the adoption of the density perspective yields results that are indistinguishable from the statistical properties generated by real jumps. In the non-Poisson case the two pictures yield totally different results. Whether or not a GME can be derived from the density approach without using implicitly or explicitly the factorization condition, is still an open issue [69].

7 Conclusions

After more than one 100 years of unquestionable successes [7], there is a general agreement that quantum mechanics affords a reliable description of the physical world. The phenomenon of quantum jumps, which can be experimentally detected, should force the physicists to extend this theory so as to turn the wave-function collapse assumption, made by the founding fathers of quantum mechanics, into a dynamical process, probably corresponding to an extremely weak random fluctuation. This dynamical process can be neglected in the absence of special cooperative effects, triggered either by the deliberate measurement act or by the fluctuation-dissipation phenomena such as Brownian motion. In this limiting case, the new theory must become identical to quantum mechanics.

We have seen that de-coherence theory, according to its advocates [7] makes the wave-function collapse assumption obsolete: the environmental fluctuations are enough to destroy quantum mechanical coherence and generate statistical properties indistinguishable from those produced by genuine wave-function collapses. All this is unquestionable, and if a disagreement exists, it rests more on philosophy than on physical facts. Thus, there is apparently no need for a new theory. However, we have seen that all this implies the assumption that the environment produce white noise, and the system of interest, in the classical limit, ordinary diffusion. As we move from normal to anomalous diffusion, the environmental fluctuations do not have the effect of forcing the system to recover the classical properties, under the form of the corresponding anomalous diffusion. This is symptom of a disease that seems to be generated by the departure from the condition of ordinary statistical mechanics. The experimental observation of quantum jumps is compatible with the Lindblad structure for a master equation [37], which, in turn, is generally regarded as being compatible with quantum mechanics, in spite of the doubt that the Markov approximation might conflict with the bath of the system of interest being rigorously quantum [38].

The phenomenon of blinking quantum dots implies the existence of a process corresponding to a GME that can derived from the experimental observation using the CTRW prescription. Is it possible to build up a GME with the same structure, using the Schrödinger's equation, and the corresponding quantum Liouville equation? This problem is still unsolved, in spite of many efforts done in this direction. The theoretical foundation of Eq. (39) is the same as the theoretical foundation of

$$\frac{\partial p(x,t)}{\partial t} = <\xi^2> \int_0^t dt' \Phi_\xi(t-t') \frac{\partial^2 p(x,t')}{\partial x^2}, \qquad (40)$$

which is supposed [67] to be a fair density description of the diffusion equation

$$\frac{dx}{dt} = \xi(t), \qquad (41)$$

with $\xi(t)$ being a dichotomous variable and $\Phi_\xi(t)$ the equilibrium correlation function of $\xi(t)$. We note that the operator term on the right hand side of Eq. (39) has the Lindblad structure, corresponding to the ordinary statistical mechanics and to the ordinary diffusion operator $\frac{\partial^2}{\partial x^2}$ appearing on the right hand side of Eq. (40). The correlation function $\Phi_\sigma(t)$ of Eq. (39) corresponds to the correlation function $\Phi_\xi(t)$ of Eq. (40). A very satisfactory fact would be to imagine that the quantum relaxation of $\Phi_\sigma(t)$ is statistically equivalent to the classical fluctuations ξ, a condition represented by

$$\Phi_\xi(t) = \Phi_\sigma(t). \qquad (42)$$

This attractive condition would imply that the quantum process of Eq. (39) is equivalent to a sequence of +'s and −'s, with the sojourn times in these two states being described by a function $\psi(t)$. In the exponential, or Poisson case, this condition of nice equivalence applies. However, when we apply anomalous statistical conditions, namely a departure of $\psi(t)$ from the exponential conditions, we see a departure from the trajectory description. In the classical case the adoption of CTRW and numerical simulation [18] yields the biscaling condition, whereas the exact solution of Eq.(40) yields mono-scaling.

The waiting time distribution $\psi(t)$ has the time asymptotic limit $1/t^\mu$, with $2 < \mu < 3$, which yields the scaling

$$\delta = \frac{4-\mu}{2}, \qquad (43)$$

if Eq. (40) holds true. The CTRW approach [18] yields a truncated Lévy distribution. The central part of this distribution has the scaling

$$\delta = \frac{1}{\mu - 1}. \qquad (44)$$

This distribution is truncated by two ballistic peaks scaling with $\delta = 1$. In the quantum case, the disagreement between the trajectory and the density perspective is even more striking, as discussed in depth in [66]. Attempts have been done at building up a GME that is exactly equivalent to the CTRW picture [54]. However, it is not yet clear if this structure can be derived from the Liouville approach. Furthermore, the GME depends on the time at which the observation process begins. Consequently, the concept itself of external perturbation is invalidated. It does not make sense to perturb the GME by means of external fields [59]. This yields a breakdown of the linear response properties that have been widely applied over the years to understand the absorption processes in condensed matter. In conclusion, all these difficulties might not be a proof of the failure of quantum mechanics to account for intermittent fluorescence in the non-Poisson case. They are, however, the symptom of a disease that can be cured by replacing the quantum with the CTRW prescriptions.

Acknowledgments

Dedicated to Professor El Naschie on the occasion of his 60th birthday. We acknowledge financial support from ARO through Grant DAAD19-02-1-0037 and from Welch Foundation through Grant B-1577.

References

[1] Ord GN, Fractals and the Quantum Classical Boundary. Chaos, Solitons and Fractals 10: 1281-1294, 1999.

[2] Nottale L, The Scale-Relativity Program. Chaos, Solitons and Fractals, 10: 459-468, 1999.

[3] El Naschie MS, On the unification of the fundamental forces and complex time in the $\mathcal{E}^{(\infty)}$ space. Chaos, Solitons and Fractals 11: 1149-1162, 2000.

[4] El Naschie MS, Time symmetry breaking, duality and cantorian spacetime. Chaos, Solitons, and Fractals 7: 499-503, 1996.

[5] H.D. Zeh, Found. Phys. 1: 69, 1970.

[6] Zurek WH, Decoherence, Einselection, and the Existential Interpretation (the Rough Guide). Phil. Trans. Roy. Soc. Lond. A 356: 1793-1820, 1998.

[7] Tegmark M, Wheeler JA, 100 Years of the Quantum, Scientific American Feb. issue 284: 68-75, 2001.

[8] Allegrini P, Giuntoli M, Grigolini P, West BJ, From knowledge, knowability and the search for objective randomness to a new vision of complexity, Chaos, Solitons and Fractal 20: 11-32, 2004.

[9] Beck C, Dynamical Foundations of Nonextensive Statistical Mechanics. Phys. Rev. Lett. 87 180601 (1-4), 2001.

[10] Grigolini P, Anomalous Diffusion, Spontaneous Localization and the Correspondence Principle, Lecture Notes in Physics, volume 457, Springer-Verlag, Berlin 1995, p. 101.

[11] Adler SL, Why decoherence has not solved the measuement problem: a response to P.W. Anderson. *Studies in History and Philosophy of Modern Physics*, 34: 135 -142, 2003.

[12] Bonci L, Roncaglia R, West BJ, and Grigolini P, Quantum Irreversibility and Chaos. Phys. Rev. Lett. 67: 2593-2596, 1991.

[13] Bonci L, Grigolini P, Morabito G, Tessieri L, Vitali D, Spontaneous localization, environment-induced decoherence and individual-system observations, Phys. Lett. A 209: 129-136, 1995.

[14] Vitali D, Tessieri L, and Grigolini P, Wave-function collapse and the quantum fluctuation-dissipation process. Phys. Rev. A 50: 967-976, 1994.

[15] Tessieri L, Vitali D. and Grigolini P., Quantum Jumps as an objective process of nature, Phys. Rev. A 51: 4404-4413, 1995.

[16] Ghirardi GC, Pearle P and Rimini A, Markov processes in Hilbert space and continuous spontaneous localization of systems of identical particles. Phys. Rev. A 42: 78-89, 1990.

[17] Paz JP, Habib S, and Zurek WH, Reduction of the wave packet: Preferred observable and de-ccherence time scale. Phys. Rev. D 47: 488-501, 1993.

[18] Allegrini A, Grigolini P, and West BJ, Dynamical approach to Lévy processes. Phys. Rev. E 54: 4760-4767, 1996.

[19] Allegrini A, Bellazzini J, Bramanti G, Ignaccolo M, Grigolini P, and Yang J, Scaling breakdown: A signature of aging. Phys. Rev. E 66: 015101 (1-4), 2002.

[20] Izrailev FM. Simple models of quantum chaos:Spectrum and eigenfunctions. Phys. Rep. 196: 299-392, 1990.

[21] Casati FG, Chirikov BV, Izrailev FM, and Ford J, in *Stochastic Behavior in Classical and Quantum Hamiltonian Systems* Vol. 93 of *Lecture Notes in Physics*, edited by Casati FG and Ford J. Berlin, Springer-Verlag; 1979.

[22] Zurek WH and Paz JP, Decoherence, Chaos and the Second Law, Phys. Rev. Lett. 72: 2508-2511, 1994.

[23] Shiokawa K and Hu BL, Decoherence, delocalization, and irreversibility in quantum chaotic systems, Phys. Rev. E 52: 2497-2509, 1995.

[24] Zaslavsky GM and Niyazov BA, Fractional kinetics and accelerator modes, Phys. Rep. 283: 73-93, 1997.

[25] Bonci L, Grigolini P. Laux A, and Roncaglia R, Anomalous diffusion and environment-induced quantum decoherence, Phys. Rev. A 54: 112-118, 1996.

[26] Roncaglia R, Bonci L, West BJ, and Grigolini P, Anomalous diffusion and the correspondence principle, Phys. Rev. E 51: 5524-5534, 1995.

[27] Stefancich M, Allegrini P, Bonci L, Grigolini P, West BJ, Anomalous diffusion and ballistic peaks: A quantum perspective, Phys. Rev. E 58: 6625-6633, 1998.

[28] Iomin A, Fishman S, and Zaslavsky GM, Quantum localization for a kicked rotor with accelerator mode islands, Phys. Rev. E 65: 036215 (1-9), 2002.

[29] Bettin R, Mannella R, West BJ. Grigolini P, Influence of the Environment on Anomalous Diffusion, Phys. Rev. E 51: 212-219, 1995.

[30] Floriani E, Mannella R, Grigolini P, Noise-induced transition from anomalous to ordinary diffusion: the crossover time as a function of noise intensity, Phys Rev E 52: 5910-5917, 1995.

[31] Grigolini P, Quantum Mechanical Irreversibility and Measurement, World Scientific, Singapore 1994.

[32] Breuer HP, Quantum jumps and entropy production. Phys. Rev. A 68: 032105 (1-7), 2003.

[33] Carusotto J and Castin Y, Condensate Statistics in One-Dimensional Interacting Bose Gases: Exact Results. Phys. Rev. Lett. 90: 030401 (1-4), 2003.

[34] Tessieri L, Wilkie J, and Cetinbas M, Exact norm-conserving stochastic time-dependent Hartree-Fock, arXiv: quant-ph/0406069.

[35] Giovannetti V, Grigolini P, Tesi G, Vitali D. Wave-function collapse and objective randomness. Phys. Lett. A 224: 31-38, 1996.

[36] Gisin N and Percival I C. The quantum-state diffusion model applied to open systems. J. Phys. A: Math. Gen. 25: 5677-5691, 1992.

[37] Lindblad G, On the generators of quantum dynamical semigroups. Communications in Mathematical Physics 48: 119-130, 1976.

[38] Rocco A, Grigolini P, The Markov approximation revisited: Inconsistency of the standard quantum Brownian motion model. Phys. Lett. A 252: 115-124, 1999.

[39] Kolmogorov AN, A metric invariant of transient dynamical systems and automorphisms in Lebesgue spaces. Doklady Academii Nauk SSSR 1958; 119: 861-864; Sinai YG, On the concept of entropy for a dynamic system. Doklady Akademii Nauk SSSR 124: 768-771, 1959.

[40] Pesin YaB, Characteristic Ljapunov exponents, and smooth ergodic theory, Akademiya Nauk SSSR i Moskovskoe Matematicheskoe Obshchestvo. Uspeckhi Matematicheskikh Nauk 132: 55-112, 977.

[41] Latora V and Baranger M, Kolmogorov-Sinai Entropy Rate versus Physical Entropy. Phys. Rev. Lett. 82: 520-523, 1999.

[42] Petrosky T and Prigogine I , Thermodynamic limit, Hilbert space and breaking of time symmetry. Chaos, Solitons and Fractals 2000; 11: 373-382; Poincaré resonances and the extension of classical dynamics. Chaos, Solitons and Fractals 7: 441-497, 1996.

[43] Grigolini P, Pala MG, Palatella L, Quantum measurement and entropy production. Phys. Lett. A 285: 49-54, 2001.

[44] Dehmelt H, Experiments with an isolated subatomic particle at rest. Rev. Mod. Phys. 62: 525-530, 1990.

[45] Bologna M, Grigolini P, Karagiorgis M, and Rosa A , Trajectory versus probability density entropy. Phys. Rev. E 64: 016223 (1-9), 2001.

[46] Manneville P, Intermittency, self-similarity and 1/f spectrum in dissipative dynamical systems, J. Physique 41: 1235-1243, 1980.

[47] Driebe DJ, *Fully Chaotic Maps and Broken Time Symmetry*. Kluwer Academic Publishers, Dordrecht, 1999.

[48] Gaspard P , Wang XJ, Sporadicity:Between periodic and chaotic dynamical behaviors. Proc. Natl. Acad. Sci. USA 85: 4591-4595, 1988.

[49] P. Allegrini, V. Benci, P. Grigolini, P. Hamilton, M. Ignaccolo, G. Menconi, L. Palatella, G. Raffaelli, N. Scafetta, M. Virgilio, J. Yang, Compression and diffusion: a joint approach to detect complexity, Chaos, Solitons and Fractals, 15: 517-535, 2003.

[50] Zwanzig R, in: Quantum Statistical Mechanics, ed. Meijer PHE, Gordon and Breach, London, 1966.

[51] Zwanzig R, in: Lectures in Theoretical Physics, Vol.3, eds. Brittin WE *et al.*, Interscience, New York, 1961.

[52] Kenkre VM, in: Springer Tracts in Modern Physics, Excitation Dynamics in Molecular Crystals and Aggregates, Springer, Berlin, 1982, p.1.

[53] Kenkre VM, Montroll EW, Shlesinger MF, Generalized Master Equations for Continuous-Time Random Walks, J. Stat. Phys. 9: 45-50, 1973.

[54] P. Allegrini P, G.Aquino G, P. Grigolini P, L. Palatella L, and A. Rosa A, Generalized Master Equation via Aging Continuous-Time Random Walk. Phys. Rev. E 68: 056123 (1-10), 2003.

[55] Bedeaux D, Lakatos-Lindenberg K, and Shuler KE, On the Relation between Master Equations and Random Walks and Their Solutions, J. Math. Phys. 12: 2116-2123, 1971.

[56] Neuhauser RG, Shimizu KT, Woo WK, Empedocles SA, and Bawendi MG, Correlation between Fluorescence Intermittency and Spectral Diffusion in Single Semicconductor Quantum Dots, Phys. Rev. Lett. 85: 3301-3304, 2000.

[57] Sugisaki M, Ren H-W, Nishi K, and Masamuto Y, Fluorescence Intermittency in Self-Assembled InP Quantum Dots 86:4883-4886, 2001.

[58] Kuno M, Fromm DP, Hamann HF, Gallagher A, and Nesbitt DJ, Nonexponential "blinking" kinetics of single CdSe quantum dots: A universal power law behavior, J. Chem. Phys. 112: 3117-3120, 2000.

[59] Sokolov IM, Blumen A and Klafter J, Dynamics of annealed systems under external fields: CTRW and the fractionak Fokker-Planck equation, Europhys. Lett. 56: 175-180, 2001.

[60] Aquino G, Bologna B, Grigolini P, West BJ, Aging and Rejuvenation with Fractional Derivatives, Phys. Rev. E 70: 036105 (1-11)2004.

[61] Aquino G, Palatella L, Grigolini P, Absorption and Emission in the non-Poisson case, Phys. Rev. Lett. 93:050601 (1-4), 2004.

[62] Bologna M, Grigolini P Pala MG, Palatella L, Decoherence, wave function collapses and non-ordinary statistical mechanics, Chaos, Solitons and Fractals 17: 601-608, 2003.

[63] Jung YJ, Barkai E, Silbey RJ, Lineshape theory and photon counting statistics for blinking quantum dots: a Lévy walk process. Chem. Phys. 284: 181-194, 2002.

[64] Brokmann X, Hermier J-P, Messin G, Desbiolles P, Bouchaud J-P, and Dahan M, Statistical Aging and Nonergodicity in the Fluoresecence of Single Nanocrystals. Phys. Rev. Lett. 90: 120601 (1-4), 2003.

[65] Cao J, Single molecule waiting time distribution function in quantum processes. J. Chem. Phys. 114: 5137-5140, 2001.

[66] Bologna M, Grigolini P, Pala MG, Palatella L, Decoherence, wave function collapses and non-ordinary statistical mechanics. Chaos, Solitons and Fractals 17: 601-608 2003.

[67] Bologna M, Grigolini P, West BJ, Strange kinetics: conflict between density and trajectory description. Chem. Phys. 284: 115-128, 2002.

[68] Allegrini P, Grigolini P, Palatella L, West BJ, Non-Poisson dichotomous noise: higher-order correlation functions and aging. Submitted to Phys. Rev. E (cond-mat/0402494).

[69] Allegrini P, Grigolini P, Palatella L, Rosa A, West BJ, Non-Poisson processes: regression to equilibrium versus equilibrium correlation function, submittted to Physica A (cond-mat/0406120).

Bohr, Bohm and Entwined Paths

by Garnet N. Ord

Abstract

Niels Bohr and David Bohm had radically different visions of the reality underlying quantum mechanics. Bohr's picture evolved into the Copenhagen interpretation and is very much part of conventional physics. Bohm's theory is observationally equivalent to Bohr's and is much closer in spirit to classical physics. Bohm's version has never become mainstream, in part because it requires that a physical particle respond to a self-generated field. How the field could be generated non-locally by the particle is not addressed by the theory. However the new model of 'Entwined Paths' shows that, at least in one dimension, a specific space-time geometry allows a single particle to construct the full Dirac propagator. In doing this, the model shows that the contrasting views of Bohr and Bohm may well be complementary.

1 Bohr

In 1900, Max Planck made the suggestion that electromagnetic energy might be quantized. The famous requirement $E = n\,h\,\nu$ allowed Planck to derive his formula for blackbody radiation. The agreement with experimental results was excellent, but the origin of the quantization condition was unknown. Planck thought that somehow, in the walls of the blackbody, the radiation oscillators were the source of the restriction. However, in 1905 Einstein used the idea of quantized energy for light to explain the photoelectric effect. In Einstein's work light consisted of particles which also exhibited wave properties. In 1913 Niels Bohr took quantization a step further when he used the requirement to explain the spectral lines of Hydrogen. From 1913-1925 'old quantum theory' developed in which the physics community searched for quantization rules that would fit experimental data. De Broglie, in his doctoral thesis was the first to propose a universal wave-particle duality in which momentum and wavelength are related through $p = h/\lambda$. In 1925 and 1926 Heisenberg and Schrödinger contributed their formulations and the modern quantum rules were born.

From its initiation with Planck's hypothesis in 1900, quantum mechanics was and is an empirical theory forced on physicists by the results of experiments. Unlike classical mechanics, where the mathematical formulation follows from an obvious ontological picture, the Schrodinger equation in the context of quantum mechanics has no obvious ontological basis. Historically, the pursuit of a realistic model underlying Schrödinger's equation was inhibited by two

important results. One was the uncertainty principle obtained by Heisenberg in 1927, ie.
$$\Delta x \, \Delta p \geq \hbar. \tag{1}$$

There were several interpretations of this at the time but the one promoted by Bohr was that a 'particle' could not simultaneously have a precise position and velocity. This was a serious impediment to constructing an ontological model for quantum mechanics. On the one hand, the particle paradigm was obviously necessary to fit in with macroscopic classical physics, and to formulate the quantum problem mathematically. On the other hand, physicists simply did not have a mathematical object that would answer the simple question "How can a particle *not* simultaneously have a precise position and momentum?" (The answer to this question had to await the development of Fractal approaches. Professor El Naschie, more than anyone else has contributed to, and championed such development.)

The uncertainty principle encouraged the rather vague but far-reaching concepts of 'wave-particle duality' and 'complementarity' of Niels Bohr. Both ideas attempted to make physical sense out of the merger of the conflicting paradigms of wave and particle. By themselves they would have been a very 'hard sell' to a physics community that was used to the real ontology of classical physics were it not for the astonishing empirical accuracy of quantum theory. There is a strong case to be made that the current dominance of the Copenhagen picture was based more on historical contingency than on physical cogency[Cushing(1994), Norris(2000)].

The second impediment to the pursuit of a realistic model for quantum mechanics was not so much the difficulty of the problem, as it was the supreme elegance of the Copenhagen 'solution'. Quantum mechanics is an empirical theory, but its current formulation is based on classical mechanics, with its strong ontology. The simple replacements $p \to -i\hbar\nabla$ and $E \to i\frac{\partial}{\partial t}$ transport you painlessly from one regime to the other. It is very easy to forget that the prescription is an analytic continuation for which there is no known physical counterpart. Generations of physicists have undoubtedly been dismayed by the vagueness of the Copenhagen interpretation, but the simplicity of the prescription, and the empirical accuracy of the theory leave little doubt that the mathematics is, for all practical purposes, correct. If the interpretation of the mathematical framework is not entirely satisfactory then, so the argument goes, this is of little interest in the practical use of the theory.

So if we deviate from practical questions, what could be wrong with Bohr's quantum mechanics? From the perspective of a realist, the theory is incomplete. The Schrödinger equation prescribes the evolution of the wavefunction in a deterministic way but, to extract information from the wavefunction, you must invoke the measurement postulates. These postulates require that a measurement reduce the wavefunction into an eigenstate of the measurement apparatus. Exactly how and when this happens in a measurement process is not specified. It is as if nature plays the game "What's the time Mr. Wolf" with us. As observers, we play the role of the wolf. When we are not looking, the children (ie. all of reality that is not being observed) move as if they are waves. As soon as we turn around to look and make an observation, the children freeze and appear as particles. In Bohr's picture, the wolf never gets to see the children move as waves, he only sees them as particles. He can devise experiments where he deduces that they move as waves, but he can never observe a transition between particle and wave.

According to a realist perspective, conventional quantum mechanics uses wave-particle duality as a vague concept to explain the peculiarities of quantum phenomena, but it has no precise model for how Nature accomplishes this duality. The result is a mathematically elegant, observationally accurate theory that allows us to *decode* Nature's behaviour, without revealing how Nature was able to *encode* the behaviour in the physical world.

2 Bohm

In 1951, David Bohm wrote the definitive text of its time on the Copenhagen interpretation of quantum mechanics[Bohm(1951)]. In 1952, he had second thoughts. He discovered that a revision of de Broglie's pilot wave theory allowed him to reproduce the empirical predictions of standard quantum mechanics without all the problems of measurement theory. His last text [Bohm and Hiley(1993)] is a lucid description of the theory which makes a very strong case for his ontological picture.

The primary difference between Bohm's picture and that of Bohr is that Bohm retains the concept of a classical particle with a well defined position and momentum. To do this Bohm writes the wavefunction in polar form $\psi = Re^{iS/\hbar}$ and substitutes this into the Schrödinger equation

$$i\hbar \frac{\partial \psi}{\partial t} = -\frac{\hbar^2}{2m}\nabla^2 \psi + V\psi \qquad (2)$$

to get

$$\frac{\partial S}{\partial t} + \frac{(\nabla S)^2}{2m} + V + Q = 0 \qquad (3)$$

and
$$\frac{\partial R^2}{\partial t} + \nabla \cdot (R^2 \frac{\nabla S}{m}) = 0 \qquad (4)$$

with
$$Q = -\frac{\hbar^2}{2m} \frac{\nabla^2 R}{R} \qquad (5)$$

Here Q is the quantum potential and (3) is just the Hamilton-Jacobi equation for a particle in a potential $V+Q$. Quantum mechanics enters this picture through the presence of Q which arises from a quantum field (ψ) produced by the particle. The particle then has an equation of motion

$$m\frac{dv}{dt} = -\nabla V - \nabla Q \qquad (6)$$

If an ensemble of identical particles is prepared with an initial distribution of particles governed by the probability density $P = R^2$, equation(4) guarantees that this distribution will remain for all time, so that the Born postulate holds for Bohm's particles.

Bohm's picture does not need the wavefunction collapse associated with the Copenhagen interpretation. The particle's trajectory is determinate and smooth, if not as straight-forward as a classical trajectory. The uncertainty principle is not due to the geometry of a single path, but to the propagation of initial conditions which supports the principle within an ensemble of systems.

Although the Bohm picture appears to be observationally equivalent to the Copenhagen picture, it has not gained wide acceptance. One reason for this is the explicit non-locality of the theory. Wherever the particle is located, it always responds to the quantum potential Q which itself 'feels' the entire accessible domain of the particle. How the particle is able to generate and respond to such a field is not part of the Bohm theory. This type of non-locality is also present in conventional quantum mechanics, it is just less obvious in that context.

3 Entwined Paths

In comparing the Bohr and Bohm pictures, the Copenhagen view is that there is no *reality* independent of the observer, and that a measurement induces either wave or particle characteristics. Bohm, on the other hand, sees an independent reality which consists of particle *and* wave. Neither prescription actually explains how Nature could implement their version of wave-particle duality. Copenhagen is silent on how measurement could cause wavefunction collapse and Bohm does not tell us how a particle can form, and respond to, its associated wavefunction and quantum potential.

Building on previous work on Fractals[El Naschie(1995), Nottale(1992), Ord(1983)], some recent work on entwined paths[Ord and Gualtieri(2002), Ord and Mann(2003)] seems to connect both views with a single mechanism involving only the geometry of space-time paths. Here we review the entwined path model.

The basic idea behind entwined paths is very simple. Classical particles moving forward in time but suffering Brownian collisions in space give rise to probability densities that evolve in time according to the diffusion equation. The Schroedinger equation is, in a sense, a reversible diffusion equation. It combines the Fractal geometry of Wiener (Brownian) paths with a specific algebra (Complex numbers) in such a way that the microscopic reversibility of individual paths is reflected in a form of macroscopic reversibility at the level of solutions of the partial differential equation. That is, when we solve Schrödinger's equation (2) we implicitly solve the complex conjugate version of the equation which also happens to be the time reversed version as well. Since it is actually a product of the form $\psi(x,t)\psi(x,-t)$ that gives us a probability density we might expect that, instead of the single Wiener paths of diffusion we should look for pairs of paths that have some features of time reversibility to them. In particular, pairs of paths have to somehow induce a special algebra into the stochastic evolution of diffusion. Entwined paths do this in the specific case of a free Dirac particle in one dimension. The Dirac algebra is induced by the geometry of entwined pairs. The geometry involves an oscillation of the pairs in space-time that produces the familiar 'zitterbewegung' of the Dirac equation. The primary contrast with conventional theory here is simply that many of the elements built into the *interpretation* of the Dirac equation (like the Dirac Sea, virtual pairs, particle spin, ...) are explicit, 'real' phenomena in entwined paths.

An Entwined Path Random Walk (EPRW) in a periodic box is sketched in Fig.1. Here the horizontal axis is space and the vertical axis is time. A particle moves in the $x-t$ plane, on a lattice, at constant speed $c = 1$. It changes directions according to a stochastic process in which the waiting time between corners and crossing points is exponential. In the figure the path is traversed forward in t along the blue path and then backwards in t along the red. An underlying space-time lattice records the number of times an entwined path passes, recording the direction in x, the direction in t, and whether or not the immediate twin is to the left or right of it. Thus the space-time lattice registers a single traversal in one of eight states. If we do not distinguish the space-time direction of traversals and merely sum over all directions, the

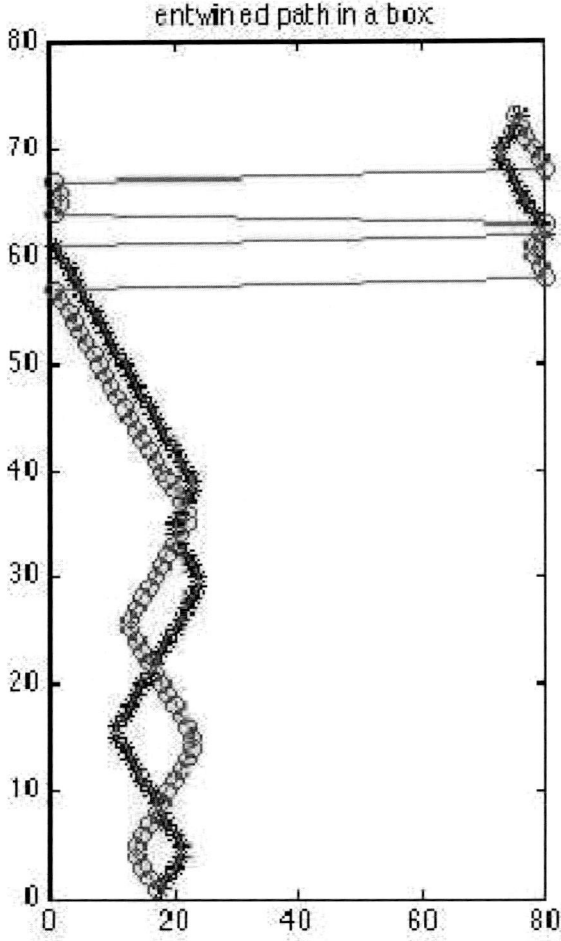

Figure 1: An EPRW. The particle starts at $t = 0$ and executes a random walk at constant speed out to beyond some return time t_r (blue path in figure). After t_r the particle reverses its direction in t and proceeds down the red path back to the origin. The 'entwined' aspect of the path is the precise nature in which the return path crosses the forward path, as described in the text. Note that between any two successive crossing points the particle and 'anti-particle' each change direction exactly once. The box has periodic boundary conditions so paths disappear at one edge but reappear at the other. This is indicated by the horizontal lines near the top of the figure.

resulting density just satisfies a discrete version of the telegraph equations. If, however, we distinguish the eight states and recognize that time reversed portions should contribute -1 instead of $+1$ we still get difference equations with a sensible continuum limit; however their qualitative behaviour is oscillatory due to the negative contributions of time-reversed trajectory segments. [Ord and Mann(2003)]

If we take the eight densities formed by many traversals of the space-time lattice and perform the subtraction of anti-particle contribution we are left with four densities which, in equilibrium (ie. after many traversals) obey the coupled equations:

$$\begin{aligned}
\phi_n^1(z) &= (1 - a\Delta t)\phi_{n-1}^1(z + c\Delta t) - a\Delta t \phi_{n-1}^2(z - c\Delta t) \\
\phi_n^2(z) &= (1 - a\Delta t)\phi_{n-1}^2(z - c\Delta t) + a\Delta t \phi_{n-1}^1(z + c\Delta t) \\
\phi_n^3(z) &= (1 - a\Delta t)\phi_{n-1}^3(z + c\Delta t) + a\Delta t \phi_{n-1}^4(z - c\Delta t) \\
\phi_n^4(z) &= (1 - a\Delta t)\phi_{n-1}^4(z - c\Delta t) - a\Delta t \phi_{n-1}^3(z + c\Delta t).
\end{aligned}$$

(7)

Here the $\phi_n^i(z)$ represent the differences in the number of red and blue trajectories that pass through a given lattice point. The equations themselves just express conservation of trajectories through lattice points. The off-diagonal terms involving $a\Delta t$ represent the corners and crossing points of the paths where particles undergo state changes by virtue of where they are going with respect to their twin. We can think of these as scattering terms.

Note that the change in signs of the scattering terms in the above equations is a direct result of the geometry of entwined pairs. If we changed all the scattering term signs to be positive we would just be adding densities and would have a discrete form of the Telegraph Equations. As it is, the system of equations is a discrete representation of the Dirac equation, one in which all amplitudes are real.

It is of interest here to note that although (7) is fully the discrete form of the Dirac equation, it also describes the equilibrium densities drawn by a single particle moving on a long entwined path. All the wave-particle features brought into quantum mechanics by formal analytic continuation is here achieved solely through the geometry of the trajectory. 'Quantum interference' in the EPRW paradigm is simply a manifestation of the fact that anti-particle trajectories can effectively erase portions of particle trajectories.

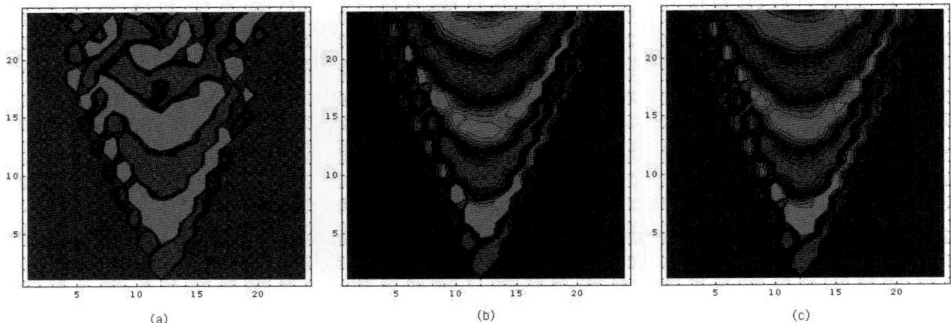

Figure 2: An entwined path draws the Dirac propagator. Pictured are contour plots of $\tilde{\phi}_3$ for respectively 10^3 10^5 and 10^7 circuits of an entwined path. In the simulation the return time is 24 lattice steps and the probability of scattering is $\alpha \Delta t = 1/2$. (c) is visually indistinguishable from the ensemble average solution obtained by solving (7) exactly. The sawtooth appearance of the light-cone is an artifact of the lattice used to store the $\tilde{\phi}$

Putting this another way, the single entwined path spends all its time constructing a 'Dirac Sea', which itself has all the properties required to provide quantum interference. The difference between EPRW and quantum mechanics is that quantum theory allows us to decode Nature's behaviour by analyzing the behaviour of a mathematical object, the wavefunction. EPRW's on the other hand show us a mechanism by which Nature could, in principle, encode that information(Figs. 2 and 3).

4 Bohr and Bohm Revisited

Since EPRW provides a completely constructive model for one dimensional quantum propagation, we have an opportunity to have a much closer look at these two very different pictures from the EPRW perspective. A preliminary study suggests that Bohm's picture of a trajectory and an associated field is absolutely right. In EPRW, the wavefunction (the source of the quantum potential Q) is drawn by entwined *pairs*. The detailed pairing of forward and backward paths is absolutely necessary for the emergence of the wavefunction from the stochastic background. On the other hand, 'measuring' a particle at a fixed time would have to prevent the reversed path from returning to the past, so a single unpaired path will remain superimposed on the wavefunction in any particle's history.

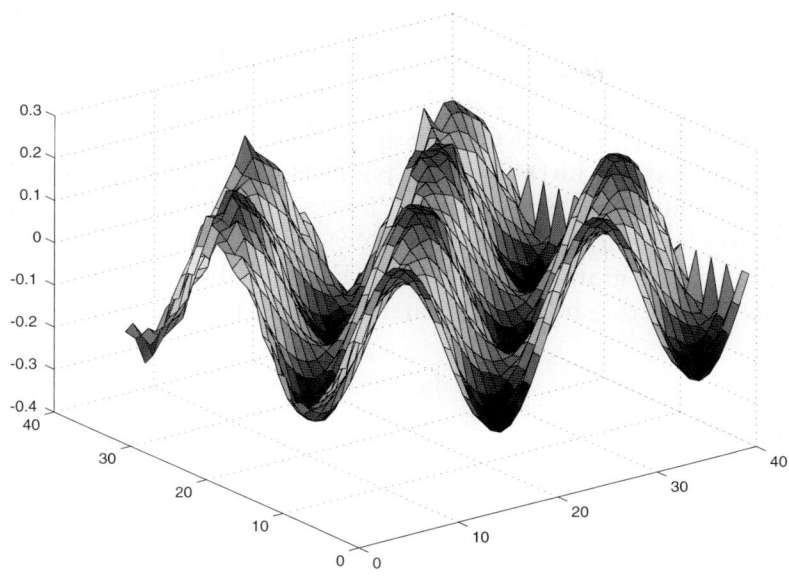

Figure 3: The second eigenstate of a particle in a periodic box as drawn by a single entwined path. Here wave-particle duality is a manifestation of the entwined geometry of the path. In the figure the x-axis is to the right and the t-axes to the left. The vertical axis measures a charge density that follows from the frequency of forward versus backwards traversals of space-time points.

Whether or not the single unpaired path is smooth and can be shown to satisfy, in the non-relativistic limit, equation(6) remains to be tested. Preliminary evidence suggests that it could.

What of Bohr and wavefunction collapse? Our current simulations contain a return time t_r beyond which an entwined path automatically reverses direction in t at the next crossing point. This time corresponds to the measurement time of Bohr. That is, for Bohr, uninterrupted propagation happens between two successive measurements. In this prescription for EPRW there is a wavefunction collapse in that on measurement, the particle is prevented from returning to its previous measurement (preparation) time. Subsequent movement in spacetime is then prescribed by the new initial conditions. Thus EPRW currently has features of both the Bohr and Bohm pictures.

Further work will be required to see if collapse is absolutely necessary, or whether Bohm's view of measurement extends to EPRW.

Acknowledgments

It is a great pleasure to dedicate this paper to Mohamed El Naschie on the occasion of his sixtieth birthday. The paper is a condensed version of a talk given at the conference 'Space Time Physics, Transfinite Mathematics and Computer Art' held in Professor ElNaschie's honour at Karlsruhe in October 2003. The author is grateful to the ZKM, Peter Weibel and Hans Diebner for the kind invitation to participate. Partial funding for this work was procided by NSERC.

References

[Cushing(1994)] J. T. Cushing, *Quantum Mechanics, Historical Contingency and Copenhagen Hegemony* (University of Chicago Press, 1994).

[Norris(2000)] C. Norris, *Quantum Theory and the Flight from Realism* (Routledge, 2000).

[Bohm(1951)] D. Bohm, *Quantum Theory* (Dover Publications, 1951).

[Bohm and Hiley(1993)] D. Bohm and B. Hiley, *The Undivided Universe* (Routledge, London, 1993)

[El Naschie(1995)] M. S. El Naschie, Chaos, Solitons and Fractals **5**, 881 (1995).

[Nottale(1992)] L. Nottale, *Fractal Space-Time and Microphysics, Towards a Theory of Scale Relativity* (Singapore: World Scientific, 1992).

[Ord(1983)] G. N. Ord, J.Phys A **16**, 1869 (1983).

[Ord and Gualtieri(2002)] G. N. Ord and J. A. Gualtieri, Phys. Rev. Lett **89** (2002).

[Ord and Mann(2003)] G. N. Ord and R. B. Mann, Phys. Rev. A **67** (2003).

On the Transition from the Classical to the Quantum Regime in Fractal Space-Time Theory

by Laurent Nottale

Abstract

In the scale-relativity theory, space-time is described as a nondifferentiable continuum and the trajectories as its geodesics. In such a space-time, the coordinates are defined as the sum of a 'classical part' that remains differentiable, and a fluctuating, 'fractal part', that is divergent and nondifferentiable. The nondifferentiable geometry has three minimal consequences, namely infinite number, fractality and irreversibility of geodesics. These three effects are accounted for by the introduction of three new terms in the total derivative acting on the 'classical part' of the coordinates. When it is written using this total derivative, Newton's equation is integrated in terms of a Schrödinger equation. Such a generalized form of the equation of dynamics is therefore both classical and quantum. In the present paper, we use this property to analyze the specific roles played by each of the individual contributions, in order to shed some light on the multiple transition between the classical and the quantum regimes.

1 Introduction

It has been discovered by Feynman [1] that the typical quantum mechanical paths (i.e., those that contribute in a main way to the path integral) are non-differentiable and fractal. Namely, Feynman has proved that, although a mean velocity can be defined, no mean mean-square velocity exists at any point, since it is given by $< v^2 > \propto \delta t^{-1}$. Although the term 'fractal' was coined only later by Mandelbrot [2], one now recognizes in this expression the behavior of a curve of fractal dimension $D_F = 2$ [3]. Based on these premises, the reverse proposal, according to which the laws of quantum physics find their very origin in the fractal geometry of space-time [4, 5, 6, 7], has been developed along three different and complementary approaches:

Ord and co-workers [8, 9, 10], extending the Feynman chessboard model, work in terms of probabilistic models, in the framework of the statistical mechanics of binary random walks.

The scale-relativity approach [6, 11, 12, 13] is founded on a nondifferentiable continuous geometry constrained by the principle of relativity (extended to scale transformations of resolutions).

El Naschie has attempted to go still one step further, and to give up also continuity. This leads him to define a 'Cantorian' space-time [7, 14], and to therefore use in a preferential way the mathematical tool of number theory. Some connections between the last two approaches have been studied in Ref. [15].

In the present short contribution, we shall consider the question of the way a physical system jumps from a classical to a quantum behavior in the scale-relativity framework. One of the goals of such a preparatory work is to study whether some difference can be found (in particular around the transition scale) between the standard quantum expectation and the fractal space-time description, which would allow one to put the new theory to the test.

2 Consequences of Nondifferentiable Geometry

Recall that the Schrödinger equation is obtained, in the scale-relativity approach, on the basis of three fundamental conditions, namely:

(1) The trajectories are in infinite number. This condition leads one to use a statistical, fluid-like description, in which the velocity $v(t)$ is replaced by a velocity field $v[x(t), t]$. The fundamental cause for this undeterminism of trajectories is the nondifferentiability of space-time, that implies its fractality [12, 13], and the subsequent identification of the trajectories with its geodesics.

(2) The trajectories are fractal curves (of fractal dimension $D_F = 2$ in the critical case that leads to standard quantum mechanics). This comes directly from the fact that the space-time itself is nondifferentiable and therefore fractal, which implies the fractality of its geodesics. This leads to generalize the concept of velocity.

Indeed, the fractal geometry implies that the velocity becomes an explicit function of the scale. Let us use a time interval δt as the scale variable. In the simplest case, the velocity will be a solution of a first order scale differential equation that reads [12, 13]

$$\frac{\partial V}{\partial (\ln \delta t)} = \beta(V) = a + bV + ... \qquad (1)$$

Such an equation can be considered as a differential fractal generator, where the coefficient b is related to the fractal dimension. Its solution is the sum of a standard, classical velocity field, and of a scale-dependent, divergent fluctuation field, $V[x(t, dt), t, dt] = v[x(t), t] + w[x(t, dt), t, dt]$, where we have identified the time interval δt with the time differential element dt. The 'fractal part' w is a fractal fluctuation of zero mean, explicitly dependent on the scale variable dt. Since we have no explicit knowledge of this fluctuation, we replace it by a stochastic variable such that $<w_k> = 0$ and (in the case $D_F = 2$ that is only considered here)

$$<w_j \times w_k> = \delta_{jk} \left(\frac{\lambda}{\delta t}\right). \tag{2}$$

We have set $c = 1$ in order to simplify the writing. By setting $dx_k = v_k dt$, $d\xi_k = w_k dt$, and $dX_k = dx_k + d\xi_k$, this relation also reads $<d\xi_j\, d\xi_k> = \lambda \delta_{jk} dt$, while $<d\xi_k> = 0$. The coefficient λ will be identified with the Compton length $\lambda = \hbar/mc$ of the particle.

(3) The invariance under the reflection transformation ($dt \leftrightarrow -dt$) is broken. Recall that this fundamental condition, that leads to a twin path process [12, 13, 16], is not set as an axiom, but is actually a consequence of the new geometry of space-time. Recall that it cannot be obtained from the mere fractality, nor from the nondifferentiability of the only trajectories. It is a consequence of the nondifferentiability of space-time itself. Indeed, the basic method of scale-relativity amounts to replace the standard physical quantities (which are defined at the limit $\delta t \to 0$ in the differentiable case) by explicitly scale dependent quantities (this is the mathematical expression of the fractality, in its general meaning). Therefore, while in its standard definition the velocity does not exist any longer since the coordinate $x(t)$ is nondifferentiable, we replace $x(t)$ by a fractal coordinate $x(t, \delta t)$, and the standard velocity is replaced by

$$v_+[x(t, \delta t), t, \delta t] = \frac{x(t + \delta t, \delta t) - x(t, \delta t)}{\delta t}, \tag{3}$$

$$v_-[x(t, \delta t), t, \delta t] = \frac{x(t, \delta t) - x(t - \delta t, \delta t)}{\delta t}. \tag{4}$$

Since one needs two points to define a velocity, there are two definitions instead of one, which are related by the reflexion on the scale variable ($\delta t \leftrightarrow -\delta t$). Note also that, contrarily to the interpretation given in our initial work on this subject ([12] Chapt. 5), these two velocities cannot be identified with Nelson's forward and backward velocities of stochastic mechanics. Indeed, the Nelson twin process [17] corresponds to a reversal of time, while here we deal with a reversal of the scale variable δt, which is an independant additional variable in the scale-relativity approach. Namely, we may write $v_-[x(t, \delta t), t, \delta t] = v_+[x(t, -\delta t), t, -\delta t]$, which means that we actually deal with a unique function, but which has no reason to be symmetrical with respect to the second variable δt. However, since the dilatation operator is $\partial/\partial \ln |\delta t|$, we need to define a twin process when jumping to logarithmic variables (which are the natural variables for describing scale transformations).

Such a twin process plays also a central role in Ord's statistical approach to the fractal space-time description [18]. In the scale-relativity approach it is a consequence of the nondifferentiable geometry itself, that supersedes fractality.

A new generalization of the velocity is involved by this fundamental symmetry breaking. We now deal with two fractal velocity fields,

$$V_+ = v_+[x(t), t] + w_+[x(t, dt), t, dt], \tag{5}$$
$$V_- = v_-[x(t), t] + w_-[x(t, dt), t, dt], \tag{6}$$

each of them decomposed in terms of a 'classical part' (v_+, v_-), which is differentiable and independent of resolution, and of a 'fractal part' (w_+, w_-), explicitly depending on the resolution interval dt and divergent at the limit $dt \to 0$.

3 The Triple Classical / Quantum Transition

Before going on, recall that these three conditions are only the minimal consequences of nondifferentiability and fractality, since far more general structures originating from nondifferentiable geometry are to be considered: they include, among other structures [19], (.) a new two-valuedness of velocity which is a consequence of nondifferentiability under the space derivative $\partial/\partial x$ and which leads to the introduction of bispinor wave functions that are solution of the Dirac equation [20], and (ii) the account of space-time dependent resolutions that lead to a geometric interpretation of the gauge fields themselves [21, 22].

In the present contribution, we consider only the simplest transition, i.e. that from the classical regime to the non-relativistic quantum mechanical regime. However, since even in this simplest case the nondifferentiability and the fractality of a space-time continuum manifest themselves under three consequences, we expect this transition to be a complicated one.

Indeed, it is a combination of the passage from a deterministic velocity on a given trajectory to a velocity field (defined on the infinity of potential trajectories), then to a fractal velocity field (i.e., that is explicitly dependent on the resolution interval), and finally to a twin fractal velocity field, namely,

$$v(t) \to v(x(t),t) \to V[x(t,dt),t,dt] \to \{V_+[x(t,dt),t,dt], V_-[x(t,dt),t,dt]\}. \tag{7}$$

Reversely, the full transition of a system from the quantum to the classical regime becomes effective only provided all of the three new properties have disappeared (either in a change of scale or in a change of the transition scale, i.e., of velocity, temperature, etc...).

4 Total Derivative in Nondifferentiable Geometry

The fundamental mathematical tool of scale-relativity amounts to include the new effects into the writing of a more complete expression for the total time derivative [12]. The three conditions imply the appearance of three additional terms:

{1} The condition (1), which tells us that the various physical quantities are functions of x and t, implies to replace d/dt by the standard 'Eulerian' total derivative

$$\frac{d}{dt} = \frac{\partial}{\partial t} + v.\nabla \ . \tag{8}$$

{2} The condition (2), which tells us that the fractal fluctuations $d\xi_k$ are now differential elements of order $1/2$, leads one to introduce terms of second order in the total derivative. Indeed, let us consider the Taylor expansion up to order two of the derivative of a physical quantity f:

$$\frac{df}{dt} = \frac{\partial f}{\partial t} + \frac{\partial f}{\partial X_k}\frac{dX_k}{dt} + \frac{1}{2}\frac{\partial^2 f}{\partial X_j \partial X_k}\frac{dX_j dX_k}{dt}. \tag{9}$$

In this expression, the differentials dX_k are the sum of a classical part dx_k and of a fractal fluctuation $d\xi_k$ of zero average. Let us consider the 'classical part' of this expression. By definition, $<dX_k> = dx_k$, so that the second term is reduced to $v.\nabla f$. Now concerning the term $dX_j dX_k/dt$, it is usually infinitesimal, but when the fractal dimension is $D_F = 2$, its 'classical part' reduces to $<d\xi_j d\xi_k>/dt$ and it is therefore finite. Thanks to Eq. (2), the last term amounts to a Laplacian, and we obtain

$$<\frac{df}{dt}> = \left(\frac{\partial}{\partial t} + v.\nabla + \frac{1}{2}\lambda\Delta\right) f \ . \tag{10}$$

In other words, the effect of the second condition is to add second order derivative terms in differential equations. Due to these terms the total derivative $<df/dt>$, despite its apparent form, should not be treated as a first order derivative, in particular as concerns the Leibniz rule for products of functions and composed functions: namely, the Leibniz rule for this derivative operator is a linear combination of the first and second order Leibniz rules (see [23] for the construction of a covariant tool allowing nevertheless to keep the form of the first order Leibniz rule, and [19] for another equivalent proposal).

{3} The condition (3), i.e. the two-valuedness of the velocity field that finds its origin in the nondifferentiability, renders the obtained description irreducible to classical, contrarily to the first two. Indeed, while the first condition is in common with the fluid description of hydrodynamics, and the second one with the description of Brownian motion, Markovian processes and random walks, there is no classical equivalent of such a twin process. The description may look similar to that of two coupled fluids, but here it is a unique 'fluid of trajectories' that is described by a twin velocity field. We have therefore suggested [12] to replace the double field (v_+, v_-) by a unique complex field

$$\tilde{V} = V - iU = \frac{v_+ + v_-}{2} - i\frac{v_+ - v_-}{2}. \qquad (11)$$

More generally, one introduces a twin classical derivative,

$$(d_+ f/dt, d_- f/dt),$$

from which we define a complex derivative operator [12]

$$\frac{d}{dt} = \frac{1}{2}\left(\frac{d_+}{dt} + \frac{d_-}{dt}\right) - \frac{i}{2}\left(\frac{d_+}{dt} - \frac{d_-}{dt}\right), \qquad (12)$$

so that we have $dx/dt = \tilde{V}$. In a more recent work [20], we have shown that the choice of complex numbers can be justified as a simplifying representation, since the relation $i^2 + 1 = 0$ allows to suppress an infinite term in the equation of motion.

Finally, when combining the effect of condition (2) (that leads to introduce second order terms in differential equations) and of condition (3) (that leads to jump from a real to a complex description), we have found [12] that the complex total derivative reads

$$\frac{d}{dt} = \frac{\partial}{\partial t} + \tilde{V}.\nabla - i\frac{\lambda}{2}\Delta. \qquad (13)$$

Finally, since $\tilde{V} = V - iU$, the three minimal consequences of the nondifferentiable and fractal geometry are expressed by the appearance in the total derivative of three additionnal terms, namely, $V.\nabla$, $-iU.\nabla$ and $-i(\lambda/2)\Delta$, so that it reads

$$\frac{d}{dt} = \frac{\partial}{\partial t} + V.\nabla - iU\nabla - i\frac{\lambda}{2}\Delta .\qquad(14)$$

5 Transition from Schrödinger to Newton Equation

As shown in many previous works [12, 13, 16, 19], if one now writes Newton's equation of dynamics (which is nothing but Einstein's equation of geodesics in the Newtonian limit when the potential is a gravitational one), in terms of the above total derivative, namely,

$$m\frac{d\tilde{V}}{dt} + \nabla\phi = 0,\qquad(15)$$

one obtains after integration a Schrödinger equation

$$frac12\lambda^2\Delta\psi + i\lambda\frac{\partial}{\partial t}\psi - \frac{\phi}{m}\psi = 0.\qquad(16)$$

In this equation $\psi = \exp(iS/S_0)$ is a mere redefinition of the action S (which is now complex since the velocity $\tilde{V}$ is complex). The Schrödinger equation is obtained provided $S_0 = m\lambda$. The constant S_0 is introduced for simple dimensional reasons: it is nothing but $\hbar$ when the theory is applied to standard quantum mechanics, but it can also be generalized in other applications (see e.g. [12] Chapt. 7, [16]). Therefore ψ acquires its status of wave function only provided it is accompanied by the Compton relation $\hbar = m\lambda$ [19].

A third form of these equations can be obtained by separating the real and imaginary parts of the Schrödinger equation and by taking as new couple of variables the real part V of the complex velocity $\tilde{V}$ and the squared modulus of the wave function $P = |\psi|^2$. We obtain a generalized Euler-Newton equation (including a 'quantum potential') and a continuity equation :

$$\left(\frac{\partial}{\partial t} + V \cdot \nabla\right)V = -\nabla\left(\frac{\phi}{m} - \frac{1}{2}\lambda^2\frac{\Delta\sqrt{P}}{\sqrt{P}}\right),\qquad(17)$$

$$\frac{\partial P}{\partial t} + \text{div}(PV) = 0.\qquad(18)$$

The full transition from the quantum to the classical regime can now be made clear. Equation 15 is both classical and quantum. When the three additional terms vanish, this is the standard deterministic equation of classical mechanics. When all of the three terms are present, its integral is the Schrödinger equation. Therefore a full study of the classical / quantum transition in the scale-relativity framework involves an analysis of the physical conditions under which these terms vanish. Such an analysis, concerning in particular its relation to decoherence and the role played by the de Broglie scale and by the thermal de Broglie scale has been initiated in previous works ([12] Sec. 5.7, [24] Sec. 4.5). We also easily verify in the last form (Eqs. 17, 18) that in the limit $\lambda \to 0$ we recover classical hydrodynamical-like equations: this corresponds to the case where only the first condition (infinity of possible trajectories described in terms of a velocity field) is fulfilled.

In what follows we shall focus on a study of the specific contribution of the two-valuedness of velocity, that leads one to jump from a real number to a complex number description.

6 A New Prime Integral of Diffusion Equations

The key condition for obtaining a genuine quantum behavior is actually the differential irreversibility condition (iii), i.e. the symmetry breaking of the reflection invariance under the transformation $(dt \leftrightarrow -dt)$. Let us demonstrate this important point by studying what happens when this condition is released.

We start from only the two first conditions, namely,

(i) infinity of trajectories;

(ii) each trajectory is fractal with $D_F = 2$.

This means that we are now describing a standard diffusion process of the Brownian motion type. The elementary displacements on each trajectory are decomposed as:
$$dX = dx + d\xi \qquad (19)$$
where
$$dx = V(x(t), t)dt, \qquad (20)$$
and
$$< d\xi^2 > = \lambda \, dt. \qquad (21)$$
i.e., $\lambda = 2\tilde{D}$ is twice the 'diffusion coefficient' in a diffusion interpretation of such a process. The velocity field $V(x(t), t)$ is now real. The effect of such a

behavior on the dynamics can be described in terms of a covariant derivative that writes:
$$\frac{D}{dt} = \frac{\partial}{\partial t} + V.\nabla + \frac{1}{2}\lambda \Delta. \qquad (22)$$
Contrarily to what happens when the differential irreversibility condition is assumed, the second order contribution $(1/2)\lambda\Delta$ is now real instead of imaginary. Namely, we jump from the quantum case to this reduced situation by making the replacement $-i\lambda \to \lambda$, and therefore $\lambda^2 \to -\lambda^2$. In terms of this total derivative operator, Newton's fundamental equation of dynamics keeps its usual form:
$$m\frac{D}{dt}V = -\nabla\phi, \qquad (23)$$
where ϕ is a potential energy. Since it preserves the form of equations, one can identify the derivative operator D/dt with a 'covariant' derivative.

One can define a Lagrange function $L(x, V, t)$ and an action S such that $dS = Ldt$, which are both real, since there is no longer any two-valuedness of the velocity vector. Let us now set
$$\varphi = e^{S/S_0}, \qquad (24)$$
where S_0 must be introduced for dimensional reasons. The function φ is a real function, that plays a role similar to that played by the complex wave function ψ. When one considers the action as a function of coordinates, one obtains $v = \nabla S/m$, so that one can replace V in Eq. (23) by
$$V = \frac{S_0}{m}\nabla \ln\varphi. \qquad (25)$$
Therefore Eq. (23) becomes
$$-S_0\left[\frac{\partial}{\partial t}(\nabla \ln\varphi) + \left(\frac{S_0}{m}(\nabla \ln\varphi.\nabla)(\nabla \ln\varphi) + \frac{\lambda}{2}\Delta(\nabla \ln\varphi)\right)\right] = \nabla\phi. \qquad (26)$$
Under the condition $S_0 = \lambda m$, this expression can be greatly simplified thanks to the identity [12]
$$2\nabla \ln\varphi.\nabla(\nabla \ln\varphi) + \Delta(\nabla \ln\varphi) = \nabla\left(\frac{\Delta\varphi}{\varphi}\right). \qquad (27)$$
We obtain:
$$\nabla\left(\lambda\frac{\partial \ln\varphi}{\partial t} + \frac{1}{2}\lambda^2\frac{\Delta\varphi}{\varphi}\right) = -\frac{\nabla\varphi}{m}. \qquad (28)$$
Therefore we find a general prime integral of the motion equations, that reads:
$$\frac{1}{2}\lambda^2\Delta\varphi + \lambda\frac{\partial\varphi}{\partial t} = \left(\frac{A(t) - \phi}{m}\right)\varphi. \qquad (29)$$

Though this equation may look like a Schrödinger equation, it has actually very different properties. The function φ is real, (while the wave function ψ was complex), the sign of the potential is reversed, and the integration function $A(t)$ does not vanish since φ has no phase. As a consequence, such an equation is not at all structuring, contrarily to the standard Schrödinger equation. It remains a diffusion equation, without counterterms allowing stationary solutions. This result gives the proof that the two-valuedness of the velocity issued from nondifferentiability is the fundamental key condition that allows to obtain a genuine quantum behavior.

Let us conclude this section by an application of this result to standard hydrodynamics. If one takes a negative value for the coefficient λ and sets $\nu = -\lambda/2$ and $m = 1$, Eq. 23 becomes the Navier-Stokes equation with a coefficient of viscosity ν. This form of the Navier-Stokes equation is obtained in every situations where $\nabla p'_r/\rho$ is a gradient (achieved when there exists an univocal link between the pressure p and the density ρ, in particular in the isentropic case where $\nabla p/\rho = \nabla w$, the enthalpy by unit of mass, see [16]). Therefore Eq. 29, that becomes

$$\nu^2 \Delta\varphi - \nu \frac{\partial\varphi}{\partial t} = \left(\frac{A(t) - \phi}{2}\right) \varphi, \tag{30}$$

is a prime integral of the Navier-Stokes equation in the case of a potential (irrotational) motion.

7 Conclusion

We have attempted in the present short contribution, to set the bases for an analysis of the classical to quantum transition in the scale-relativity approach. The transition is a complicated one, since it involves the combined vanishing of three conditions that can be deduced from the nondifferentiability of the space-time continuum: nondeterminism of trajectories, fractality ($D_F = 2$), and two-valuedness of velocity. These conditions manifest themselves by the appearance of three additional terms in the total derivative with respect to time. The term $V.\nabla$ is a fluid-like term that comes from the loss of determinism; the term $-iU.\nabla$ comes from the combination of nondeterminism and two-valuedness (described in terms of complex numbers), while the second order derivative term $-i(\lambda/2)\Delta$ comes from the combination of fractality and velocity two-valuedness.

We have particularly studied in this paper the influence of the velocity two-valuedness and of its representation in terms of complex numbers. We have shown that, in case this condition is not fulfilled, one still obtains a Schrödinger-like equation, but which is now real and in which the signs of the various terms are changed. As a consequence such an equation is far from structuring: on the contrary, it is describing a purely diffusing behavior. This demonstrates that the origin of the structuring behavior of quantum mechanics (described in terms of well-defined probability density distributions) is to be found in the doubling of the velocity field, whose origin is attributed in the scale relativity approach to the loss of the space-time differentiability.

Acknowledgments

We acknowledge the editors of this Festschrift for their kind invitation to contribute.

References

[1] R. P. Feynman and A. R. Hibbs, *Quantum Mechanics and Path Integrals* (MacGraw-Hill, New York, 1965)

[2] B. Mandelbrot, *Les Objets Fractals* (Flammarion, Paris, 1975)

[3] L.F. Abbott and M.B. Wise, 1981, Am. J. Phys. 49, 37

[4] G. N. Ord, J. Phys. A: Math. Gen. **16**, 1869 (1983)

[5] L. Nottale and J. Schneider, J. Math. Phys. **25**, 1296 (1984)

[6] L. Nottale, Int. J. Mod. Phys. A **4**, 5047 (1989)

[7] M. S. El Naschie, Chaos, Solitons & Fractals **2**, 211 (1992)

[8] D. G. C. McKeon and G. N. Ord, Phys. Rev. Lett. **69**, 3 (1992)

[9] G.N. Ord, Chaos, Solitons & Fractals **7**, 821 (1996)

[10] G.N. Ord and J.A. Galtieri, Phys. Rev. Lett. 89, 250403 (2002)

[11] L. Nottale, Int. J. Mod. Phys. A**7**, 4899 (1992)

[12] L. Nottale, *Fractal Space-Time and Microphysics: Towards a Theory of Scale Relativity* (World Scientific, Singapore, 1993)

[13] L. Nottale, Chaos, Solitons & Fractals **7**, 877 (1996)

[14] M. S. El Naschie, in "Quantum Mechanics, Diffusion and Chaotic Fractals", Eds. M.S. El Naschie, O.E. Rssler and I. Prigogine (Pergamon, 1995), p. 191

[15] M. S. El Naschie, Chaos, Solitons & Fractals **11**, 2391 (2000)

[16] L. Nottale, Astron. Astrophys. 327, 867 (1997)

[17] E. Nelson, Phys. Rev. **150**, 1079 (1966)

[18] G.N. Ord and A.S. Deakin, Phys. Rev. A 54, 3772 (1996)

[19] L. Nottale, in "Computing Anticipatory Systems. CASYS'03 - Sixth International Conference" (Liége, Belgique, 11-16 Aug 2003), Daniel M. Dubois Editor, American Institute of Physics Conference Proceedings 718, 68 (2004)

[20] M.N. Célérier & L. Nottale, J. Phys. A: math. Gen. 37, 931(2004)

[21] L. Nottale, in *Relativity in General*, (Spanish Relativity Meeting 93), edited by J. Diaz Alonso and M. Lorente Paramo, (Editions Frontières, Paris, 1994) p. 121

[22] L. Nottale., M.N. Célérier, T. Lehner, G.R.G., accepted (2004) (eprint hep-th/0307093)

[23] J. C. Pissondes, J. Phys. A: Math. Gen. 32, 2871 (1999)

[24] L. Nottale, in "Quantum Mechanics, Diffusion and Chaotic Fractals", Eds. M.S. El Naschie, O.E. Rössler and I. Prigogine (Pergamon, 1995), p. 51

Mohamed El Naschie's $\epsilon^{(\infty)}$ Cantorian space-time and its consequences in cosmology

by Gerardo Iovane

Abstract

In this paper we introduce Mohamed El Naschie's $\epsilon^{(\infty)}$ Cantorian space-time in connection with stochastic self-similar processes to give a possible explanation of the segregation of the Universe at fixed scale; then we summarize some relevant consequences.

1 Introduction

Nature shows us structures with scaling rules, where clustering properties from cosmological to nuclear objects reveals a form of hierarchy. Mohamed El Naschie has realized a very wide scientific production in few hundred papers on $\epsilon^{(\infty)}$ Cantorian space-time. In the last 20 years he has made the theory, considered many and relevant consequences in fundamental physics, studied the application to many relevant examples. Most relevant results were in theoretical physics and high energy physics. Nevertheless his point of view is more general and applicable in many other contexts such as cosmology, mathematical physics, analysis, and applied sciences. In this contribution I wish to show some consequences in cosmology.

In a previous paper[4], the consequences of a stochastic, self-similar, fractal model of the universe was compared with observations. In particular, it was demonstrated that the observed segregated Universe is the result of a fundamental self-similar law, which generalizes the Compton wavelength relation, $R(N) = (h/Mc)N^\phi$, where R is the radius of the structures, h is the Planck constant, M is the total Mass of the self-gravitating system, c the light speed, N the number of nucleons within the structures, and $\phi = \frac{\sqrt{5}-1}{2}$ is the Golden Mean value [1]. As noted by Mohamed El Naschie this expression agrees with the Golden Mean and with the gross law of Fibonacci and Lucas [2],[3].

In [4] G.Iovane showed the relevant consequences of a Stochastic Self-Similar and Fractal Universe. Starting from an universal scaling law, the author showed its agreement with the well–known Random Walk equation or Brownian motion relation that was used by Eddington [5], [6]. Consequently, he arrived at a self-similar Universe. It appears that the Universe has a memory of its quantum origin as suggested by R.Penrose with respect to quasi-crystal [7]. Particularly, the model is related to Penrose tiling and thus to $\varepsilon^{(\infty)}$ theory (Cantorian space-time theory) as proposed by El Naschie [8],[9] as well as with Connes Noncommutative Geometry [10]. In [1] the authors presented a descriptive model of segregated universe, then considered a dynamical model to explain the results and to give the evolution of the structures [11]. In [12], waveguiding and mirroring effects are considered with respect to the large scale structure of the Universe.

In [13] it was considered how a Cantorian space could explain some relevant stochastic and quantum processes, if the space acts as an harmonic oscillating support, such as happen in Nature. In other word, the vision is that an apparent indetermination, linked with a fractal support rather than a continuous one, can produce an indetermination on the motion of a physical object, which is explained via a stochastic or quantum process. This means that a quantum process, in some cases, could be explained as a classical one, but on a noncontinuous and fractal support. Consequently, an external observer looking at the motion of a particle under a fixed solicitation can measure an unusual behaviour with respect to a continuous material support, that is obvious with respect to the knowledge of the fractal support behaviour. In this case, he can make the hypothesis of an indetermination or a stochasticity in the process (motion), while there is just really only ignorance with respect to the support on which the motion take place.

2 E-infinity Cantorian space and stochastic self-similar random processes

In descriptive set theory and the theory of polish spaces it is shown that [14].
Definition 1. *When a space $A^{\mathbb{N}}$ is viewed as the product of infinitely many copies of A with discrete topology and is completely metrizable and if A is countable, then the space is said to be polish.*

In particular, when $A = \{0, 1\}$, $|A| = 2$, then we call $\mathcal{C} = \in^{\mathbb{N}}$ Cantor space. For A^{-1} defined in an interval $A^{-1} \subset]0,1[$ then $C_F = A^{\mathbb{N}}$ is called a fuzzy Cantor space. If $|A^{-1}| = (\sqrt{5} - 1)/2$ and $N = n - 1$, where $-\infty \leq n \leq \infty$, then $C_F = \varepsilon^{(n)}$ is the E-infinity Cantorian space. Mohamed El Naschie in [15] showed the relationship between the Cantor space $\mathcal{C}$ and $\varepsilon^{(\infty)}$. As he reports: " the relationship comes from the cardinality problem of a Borel set in polish spaces Thus we call a subset of ε topological space a Cantor set if it is homeomorphic to the Cantor space".
From the other hand we have the following scenario with respect to stochastic self similar procsses.

Let $\Re$ be real space and $\gamma_r \in \Re^+$, then we define a self-similar (ss) random process for every $r > 0$,

$$X(s) \stackrel{d}{=} \gamma_r X(rs), \qquad \text{with} \qquad s \in \Re, \tag{1}$$

where $\stackrel{d}{=}$ denotes equality as distributions [16].

The relation (1) is invariant under the group of positive affine transformations,

$$X \to \gamma X, \qquad s \to rs, \qquad \gamma_r > 0. \tag{2}$$

Since γ_r satisfies the properties

$$\gamma_{r_1 r_2} = \gamma_{r_1} \gamma_{r_2}, \qquad \forall r_1, r_2 > 0, \tag{3}$$

$$\gamma_1 = 1,$$

then it must have the form

$$\gamma_r = r^{-\delta}, \qquad \text{with} \qquad \delta \in \Re. \tag{4}$$

Thanks to (4) the relation (1) becomes

$$X(s) \stackrel{d}{=} r^{-\delta} X(rs), \qquad \text{with} \qquad s \in \Re. \tag{5}$$

When a process satisfies (1) or (5), it is said to be self-similar or δ−self-similar.

A generalization of self-similar random process is obtained by replacing the deterministic scaling factor $\gamma_r = r^{-\delta}$ in (1) or (5) with a random variable $\widetilde{\gamma}_r \in \Re_0^+$. This variable is independent of the process to which such a variable is multiplied. Then eq.(5) becomes

$$X(s) \stackrel{d}{=} \widetilde{\gamma}_r X(rs), \qquad \text{with} \qquad s \in \Re. \tag{6}$$

D.Veneziano demonstrated in [17] that $\widetilde{\gamma}_r$ can also be written as $\gamma_r = r^{-\widetilde{\delta}}$ with $\widetilde{\delta}$ real random variable. Then, these kinds of processes, called stochastic self-similar (sss) random processes and the previous ones (ss), can be treated in the same theory. Gupta and Waymire showed that for $0 < r \leq 1$ the sss processes are dilations, while for $r > 1$ the sss processes are contractions [18],[19].

In [17] the author proved the following relevant theorem: if $\widetilde{\delta}_{r_1} \stackrel{d}{=} \widetilde{\delta}_{r_2}$ for some $r_1 \neq r_2$, then $\widetilde{\delta}$ must be a deterministic constant δ. Then, one can treat ss and sss random processes in a unique scheme.

Moreover, the author gives many relevant properties and generalizations to a d-dimensional space in the same paper, but we are not going to consider these properties because they do not fit the objectives of our paper (for more details see [17]).

3 The physical scenario and the unification of the fundamental interactions in $\varepsilon^{(\infty)}$ Cantorian Space-Time

Following El Naschie [5], let us start from Plank's fundamental energy quantization relation

$$E_P = h\nu = h\frac{c}{\lambda_C}, \tag{7}$$

where E is the energy, h the Plank constant, ν the frequencies, c the speed of light and λ_C is the Compton wavelength. On the other hand, the Einstein mass-energy equation of special relativity is

$$E_E = mc^2, \tag{8}$$

where m is the mass. Consequently, we obtain

$$\lambda_C = \frac{h}{mc}, \tag{9}$$

this is the well-known Compton wavelength relation [20].

Iovane in [1] demonstrated that (9) is a special case of the self-similar law

$$R(N) = \frac{h}{mc}N^{1+\phi} = \frac{h}{m_n c}N^\phi, \tag{10}$$

where R is a scale length (for example the dimension of the Universe, the length of a super cluster, the length of a galaxy, the scale length of a man, of a cell, or the nucleons radius), m is the mass, N is the number of nucleons inside the structure, m_n is the mass of a nucleon (roughly speaking the mass of a proton or neutron), ϕ is the Golden mean value.

From the (7) and (10) it is easy to find the following general expression, which links the energy at macroscopic scale with the microscopic one, thanks to a self-similar relation, that takes into account the numerical density, too

$$E_{E,N}(N) = E_P N^{1+\phi}, \iff E_E = \widetilde{E}_{P,N}(N), \tag{11}$$

where

$$\widetilde{E}_{P,N}(N) = h\widetilde{v}, \tag{12}$$

with

$$\widetilde{v}(N) = v N^{-(1+\phi)}. \tag{13}$$

Consequently, It exists a characteristic frequency of the scale for each a system. It is linked with its mass and extension. The following theorems are obtained.

Theorem 1. *The structures of the Universe appear as if they were a classically self-similar random process at all astrophysical scales. The characteristic scale length has a self-similar expression*

$$R(N) = \frac{h}{Mc}N^{1+\phi} = \frac{h}{m_n c}N^{\phi},$$

where the mass M is the mass of the structure, m_n is the mass of a nucleon, N is the number of nucleons in to the structure and ϕ is the Golden Mean value.

In terms of Planckian quantities the scale length can be recast in

$$R_P(N) = \frac{l_P}{m_P}\sqrt{\frac{\hbar c}{G}} N^{(1+\phi)}.$$

The previous expression reflects the quantum memory of the Universe at all scales, which appears as hierarchy in the clustering properties.

Theorem 2. *The mass and the extension of a body are connected with its quantum properties, through the relation*

$$E_{E,N}(N) = E_P N^{1+\phi},$$

that links Plank's energy and Einstein's one.
The quantum memory is reflected at all scales and it manifests itself through a clusterization principle of the mass and extension of the body.

As suggested by El Naschie in [5], the Eq. (7) may be viewed as a potential of a force F at height $\overline{H}$. If we assume $\overline{H} = \lambda$, it follows

$$-V = E_P = F\overline{H} = \left(\frac{hc}{\lambda^2}\right)\lambda, \qquad (14)$$

and so the force F results

$$F = -\frac{dV}{d\lambda} = \frac{hc}{\lambda^2}. \qquad (15)$$

The Eq. (5) in terms of (4) becomes

$$E_{E,N}(N) = \frac{hc}{R(N)} N^{2(1+\phi)}. \qquad (16)$$

Consequently, we can easily obtain the generalization of (15), that gives us

$$F = -\frac{dE_{E,N}(N)}{dR} = E_{E,N}(N)/R(N). \qquad (17)$$

In other words, our expression of $E_{E,N}(N)$ preserves the classical approach to obtain a conservative force from a potential. Consequently, the work to move a body of a distance R, as in the classical case, is

$$L = E_{E,N} = FR. \tag{18}$$

The previous energy can be also considered as the binding energy or the intrinsic energy of the system.

As showed by El Naschie, a similar result can be obtained for the Yukawa theory too.

4 Conclusions

In this paper we have considered the link between Mohamed El Naschie's $\epsilon^{(\infty)}$ Cantorian space-time and the stochastic self-similar random processes. Then we have considered the consequences in cosmology. In particular, it was demonstrated that the observed segregated Universe is the result of a fundamental self-similar law, which generalizes the Compton wavelength relation, $R(N) = (h/Mc)N^\phi$. As noted by Mohamed El Naschie this expression agree with the Golden Mean and with the gross law of Fibonacci and Lucas. Moreover it appears that the Universe has a memory of its quantum origin like as suggested by R.Penrose with respect to quasi-crystal. Particularly, it is related to Penrose tiling and thus to $\varepsilon^{(\infty)}$ theory (Cantorian spacetime theory) as proposed by M.S. El Naschie as well as in A.Connes Noncommutative Geometry.

References

[1] G.Iovane, E.Laserra, F.S.Tortoriello, Stochastic Self-Similar and Fractal Universe, Chaos, Solitons and Fractals, 20, 3, 415, 2004.

[2] T.A. Cook, The curves of life. Originally published by Constable and Company, London 1914; Reprinted by Dover Publications, New York. See also: M.S. El Naschie, Multidimensional Cantor-like Sets ergodic behaviour, Speculation in Science & Technology, 15,(2), 138-142, 1992.

[3] S. Vajda, Fibonacci and Lucas Numbers and the Golden Section, J.Wiley, New York, 1989.

[4] G.Iovane, Varying G, accelerating Universe, and other relevant consequences of a Stochastic Self-Similar and Fractal Universe, Chaos, Solitons and Fractals, 20, 4, 657, 2004.

[5] M.S. El Naschie, On the unification of the fundamental forces and complex time in the $\varepsilon^{(\infty)}$ space, Chaos, Solitons and Fractals, 11, 1149-1162, 2000.

[6] B.J.Sidharth, Chaos Solitons Fractals 11, 2155, 2000; Chaos Solitons Fractals 12, 795, 2001.

[7] R.Penrose, The Emperor's New Mind, Oxford University Press, 1989.

[8] M.S. El Naschie, On the uncertainty of Cantorian geometry and the two slit experiment, Chaos, Solitons and Fractals, 9, 3, 517-529, 1998.

[9] M.S. El Naschie, Penrose universe and Cantorian spacetime as a model for noncommutative quantum geometry, Chaos, Solitons and Fractals, 9, 931-933, 1998.

[10] A. Connes, Noncommutative Geometry, Academic Press, New York, 1994.

[11] G.Iovane, E.Laserra and P.Giordano, Fractal Cantorian structures with spatial pseudo-spherical symmetry for a possible description of the actual segregated universe as a consequence of its primordial fluctuations, Chaos, Solitons and Fractals, 22, 4, 521, 2004.

[12] G.Iovane, Waveguiding and Mirroring Effects in Stochastic Self-Similar and Fractal Universe, Chaos, Solitons and Fractals, 23, 3, 691, 2004.

[13] G.Iovane, P.Giordano, S.Salerno, Dynamical Systems on El Naschie's $\epsilon^{(\infty)}$ Cantorian space-time, Chaos, Solitons and Fractals, 24, 2, 423, 2005.

[14] T.Jech, Set Theory, Berlin: Springer, 2003.

[15] M.S. El Naschie, Quantum gravity, Clifford algebras, fuzzy set theory and fundamental constants of nature, Chaos, Solitons and Fractals, 20, 437, 2004.

[16] W.Vervaat, Bull.Int.Statist.Inst. 52, 199, 1987.

[17] D.Veneziano, Fractals, Vol.7, No.1, 59, 1999.

[18] V.K.Gupta and E.C.Waymire, J.Geophys. Res. 95, 1999, 1990.

[19] V.K.Gupta et al., Water Resour.Res. 95, 3405, 1994.

[20] M.Alonso, E.Finn, Fundamental university physics III. Quantum Mechanics and statistical physics, New York:Addison-Wesley, 1968.

Needle People and Pancake People: The Gulliver Effect

by Otto E. Rossler

Abstract

The Gulliver effect is a vertical size change implicit in the equivalence principle. Unlike FitzGerald contraction, it can be both positive and negative. Moreover, it is reciprocal - while one party appears expanded to the other, the other appears contracted to the former: needle formation implies pancake formation and vice versa. Computer simulations are encouraged. The connection to general relativity is not yet understood.

1 Introduction

We are all needle people without our noticing. This is easy to demonstrate: All you need is a single-pulse pocket laser, two pocket mirrors, a fast counter and an exact timer. Leave one of the mirrors flat on the ground and climb a high tower. Measure the return time of the light pulse sent down and reflected back (the second mirror if mounted upstairs and the counter make it possible to measure the cumulative duration of many consecutive down-up roundtrips so that a less expensive clock suffices). Then repeat the experiment from the ground using the mirror left upstairs. The two measured times (down-up and up-down) will be different.

Since the speed of light is locally the same upstairs and downstairs, the experiment proves that the people downstairs are taller by a factor of about $1 + 10^{-14}$ (if the tower is 200 m high) according to the redshift measurement of Pound and Rebka [1]. Amazingly, the suggested experiment has never been done (Gerhard Schfer, personal communication 2001), even though it can be done [2].

The implied relative elongation or shrinking is the Gulliver effect, cf. [3]. The effect is not necessarily tiny. When leaving the mirror on the surface of a neutron star, for example, a size increase of about 20 percent of any nuclear-reactions based life form living there is implicit. This is because the aspect ratio (ratio between height and width) of the needle people is given by the longitudinal Doppler factor (gravitational redshift factor), which is about 120 percent on neutron stars.

A mirror left on the horizon of a black hole (and kept afloat), in contrast, would be subject to an infinite redshift and hence acquire an infinite aspect ratio. Conversely, to the people floating on the horizon, we (the upper-world people) would be infinitely flat and infinitely close while our width would be unchanged. Thus infinitely strong Gulliver effects are in principle realizable. What is so special about them?

2 Understanding the Gulliver Effect

The Gulliver effect is not a "contact effect" like the FitzGerald contraction [4]. It therefore is not confined to being negative (contractive). Moreover unlike the FitzGerald contraction which is symmetric (either of two objects in relative motion is contracted for the other), the Gulliver effect is reciprocal: The one side sees the other expanded while the latter sees the first contracted. This at least is what holds true in the context of the equivalence principle (that is, in the context of acceleration) where the effect is the most readily demonstrated.

The effect is a consequence of the fact (known since Einstein's paper in 1907) that any time interval downstairs in an accelerating rocket, is elongated by the gravitational redshift factor. Hereby the words "any time interval downstairs" include up-down return times measured downstairs. Since the same reflected light signals are involved for both sides and since the speed of light is locally the same upstairs and downstairs, it indeed follows that the *up-down* light distance is shorter than the *down-up* light distance by the Doppler factor [2]. Taking into account that the lateral distances are unchanged under Lorentz transformations, it follows that the people upstairs are vertically contracted (into pancakes) for the people downstairs while the people downstairs are vertically expanded (into needles) for the people upstairs.

3 Spatial Doppler Effect

The Gulliver effect was introduced above as a new consequence of the longitudinal Doppler effect. A better way to understand it, is to say that it represents a new Doppler effect of its own, a "spatial Doppler effect." For just as light emitted from the end points of a *temporal* interval on a departing point-shaped object is Doppler-expanded in its temporal distance on arrival at a point in the rest frame ("temporal shadow"), so light emitted from the end points of a *spatial* interval on a departing longitudinal object is Doppler-expanded in its spatial distance on arrival at a line in the rest frame ("spatial shadow").

But mere shadows do not represent reality? This, surprisingly, is not always true. Einstein discovered the reality of the temporal shadow. This is the most astounding discovery of his (the so-called twin paradox). He saw that when the object with the distorted temporal shadow is brought back, it turns out that the shadow (the distorted clock rate) has been telling the truth all along: The twin brought back (or hauled back up again, respectively, in case of the equivalence principle) is younger!

The same thing unexpectedly happens in the present spatial case: Any object whose spatial shadow is enlarged is enlarged itself in effect. The first proof of the "shadow paradox" was already presented above: The shadow people need fewer seconds to cover a given longitudinal distance with light. Hence they must have been larger while living downstairs.

The second proof of the "shadow paradox" invokes angular-momentum conservation ($L = \omega r^2 m =$ const.). If the changed clock rate ω is real in accord with the clock paradox of the equivalence principle, the size r cannot remain constant in the product L if m (the mass) is unchanged. So much for the good news. Bbut the change of r cannot be linear in that case. If $r\omega =$ const. as claimed for the shadow downstairs, mass m must go down in parallel with the frequency ω ($rm =$ const.). The constancy of rm follows from quantum mechanics [5].

4 Discussion

Special relativity contradicts common sense in a new way. Unilaterally watched objects in motion are distorted, not only in time but also in space. These ghostly changes are mere projection effects. Mysteriously, they tend to get confirmed when checked. The 2 clock paradoxes of Einstein were shown to possess analogs in the spatial domain. The Gulliver effect arises as a spatial analog of the temporal twin paradox. A greater symmetry of special relativity is thereby achieved. What are the consequences?

The size change of the spatial Doppler effect introduces a similar gradation into spacetime as is familiar from the temporal Doppler effect. The clock-rate change is an element of general relativity, as is well known. How about the size change?

This question cannot yet be answered. When time slows down maximally near a black hole, for example, size must go up maximally in the vertical direction, as we saw. It appears that this effect is not part of general relativity as we know it. The new symmetry achieved in special relativity and equivalence principle would thereby wreck havoc in the next-higher building block. This casts grave doubts on its existence.

One dreadful consequence usually does not come alone. How about energy conservation? As we saw, mass (and hence energy) is not conserved downstairs. This fact is, of course, well known for photons [1]. However, a cyclic light transmission between upstairs and downstairs does preserve energy (and mass). The new mass change may, therefore, be acceptable.

To conclude, a spatial clock paradox has been described. Size is equally malleable in relativity as time, both contraction-wise and expansion-wise. A second mere projection effect after the clock paradox surprisingly proves real. Needles and pancakes become part of the nonlinear science of special relativity including acceleration, just as noodles and pancakes are part of the nonlinear science of Poincaré [6]. Both subfields of chaos, solitons and fractals seem to grow together. The beautiful fractal-dust picture in infinite-dimensional space, used by El Naschie to derive the dimensionality of spacetime and the mass ratios of elementary particles [7], stands like an attractor behind both subfields.

Acknowledgments

Dedicated to Mohamed El Naschie on the occasion of his 60th birthday. I thank Lydia and Garnett for stimulation. I also thank David Finkelstein and George Galeczki for discussions. Dieter Fröhlich, Normann Kleiner, Martin Pfaff and Heinrich Kuypers contributed. For J.O.R.

References

[1] Pound RV, Rebka GA. Apparent weight of photons. Phys Rev Lett 1960;4:337-41.

[2] Rossler OE, Kuypers, H, Parisi. Gravitational slowing down of clocks implies proportional size increase. Lecture Notes in Physics 1998; 503:370-72.

[3] Rossler OE. The Gulliver effect. In: Diebner HH, Druckrey T, Weibel P, editors. Sciences of the Interface (Festschrift). Tübingen: Genista; 2001, p. 1-6.

[4] FitzGerald GF. The ether and the earth's atmosphere. Science 1889;13:390. Reprinted in: Pais A. Subtle Is the Lord, The Life and the Sciences of Albert Einstein. Oxford: Clarendon; 1982, p. 122.

[5] Kuypers H, Rossler OE, Bosetti P. Matterwave-Doppler effect, A new implication of Planck's old formula $E = h\nu$ (in German). WechselWirkung 2003;25(2):26-27.

[6] Rossler OE. Chaos and bijections across dimensions. In: Holmes PJ, editor. New Approaches to Nonlinear Problems in Dynamics. Philadelphia: Siam; 1980, p. 477-86.

[7] El Naschie MS. On a general theory of quantum gravity. In: Diebner HH, Druckrey T, Weibel P, editors. Sciences of the Interface (Festschrift). Tübingen: Genista; 2001, p. 72-75.

Cosmic Computation

by David Ritz Finkelstein

Abstract

We describe the cosmos as a process composed of elementary atomic processes on a quantum set of quantum bits. We model space-time, field, and Higgs field by a condensation of the cosmos at the second level of analysis (bits of bits).

1 The Physics Process

For me physics is not the search for a final all-embracing theory but the ongoing revision of theory to fit the never-ending surprises of experiment; a process, not a product. Therefore one should pay attention to the physics process itself. When we do, we see, among other important evolutionary processes, one sequence of reforms, relativizations, and renunciations that make physics more regular, non-commutative, operational, relativistic, atomistic, and simple. Each of these reforms is associated with a physical constant that gives the strength of the non-commutativity, and each relativizes some false absolute. Some of these physical constants and relativized absolutes are:

k: kinetic theory	~~caloric~~
c: special relativity	~~time-~~
R: de Sitter space-time	~~translation~~
G: general relativity	~~acceleration-~~
h: quantum theory	~~state~~
e: gauge electrodynamics	~~electromagnetic phase~~
g: gauge chromodynamics	~~chromodynamic phase~~
W: gauge electroweak dynamics; W =W mass	~~electric axis~~
h', h'': Regular quantum theory	~~quantum phase~~
...	

I have optimistically tacked on the reform that my colleagues and I am currently attempting and describe here, with two companions h', h'' to the quantum constant h. Many of these, perhaps all, involve one process of Lie algebra (or group) regularization. I use this term for a process of Segal [17] that he did not trouble to name. To regularize an algebra that is neither simple nor stable one inserts a small physical constant p called the regularizing parameter into its multiplication table, changing it slightly so that it becomes simple and stable. The regular theory reduces to the singular theory when the constant p goes to 0. A famous special case of the inverse process is called group contraction [12]. I extend that term to the general case.

Contracting a theory is a straightforward mathematical process reducing its domain of validity. Regularizing a theory in a way that enlarges its validity domain is an ill-posed inverse problem, which may have several solutions or none. We concentrate on group regularization here.

Why so many? They are all simplifications. Are we approaching one final simple theory in the sky?

Maybe; but all the ancient groups are borderline cases. They lie on the boundaries between two regions filled with isomorphic stable groups, and they can go either way at the least provocation, the smallest improvement in experimental precision [17]:

$$\text{Compound, unstable} \longrightarrow \frac{\text{Simple, stable}}{\text{Simple, stable}}$$

Unstable theories naturally have a higher mortality rate than stable ones. The process may be Darwinian, not teleological, and the selector may be survival, not an ultimate perfection.

Typically the regular theory explains previously mysterious or even unperceived assumptions of the singular theory, while the regular theory seem magical or self-inconsistent from the singular viewpoint. At a critical early phase in the development of mathematical physics we acquired a deep prejudice for the singular against the regular, so that Euclidean geometry was taken to be more fundamental than spherical. Now it seems that all the infinities of physics come from its singular groups. Nature puts no infinities into our theories. We put them there ourselves, by fixating on the singular over the regular. Each relativization creates a vanguard of physicists who sally into the regular domain and a rear guard who keeps its base in the singular domain. The conservatives in the rear guard stabilize physics against over-imaginative theorizing, while the reformers in the van begin the next relativization. Here we venture into the van.

2 A Logic of Process

We will represent the cosmos as a kind of computer, a "logical engine." For one relatively early expression of a purely digital (discrete) world see the

Vaisheshiko Sutra (circa 100 AD). For a later model see the discussion of the Kalam in Maimonides "Guide of the Perplexed" circa 1200; the concept he discusses goes back to the Islamic Mutakalimun of Bagdad circa 800 AD and may have been influenced by the Hindu sutras. Since the writers denied causal determinism on theological grounds, they might even be said to have some aspect of quantum theory built into their model, if one is willing to stretch the concept. The philosophical idea of state-relativity, another element of the quantum theory, is also old.

The first step is to use the right logic. We take up its philosophy in this section and the algebra in the next.

As Bohr, Von Neumann, and others agreed at the start, classical logic does not work in the quantum domain. The h regularization strikes at the most cherished absolute so far, the absolute Truth as summed up in the absolute State. Classical logic is totalitarian: The State determines the value of all variables. The State is not only the Truth, it is the whole Truth. The State is then absolute in that all ideal observers are required to agree on its value. Classical ($=$ c) physics attributes such an all-determining State variable to each system. The h regularization to quantum ($=$ q) physics relativizes the state and so dethrones the absolute State and absolute Truth as the c regularization does absolute Time and absolute Mass.

Those who attribute to a quantum system an absolute state, even one that collapses on observation, are therefore still in the rear guard of the h reform. Where equations of classical physics make statements about the state of being, those of quantum physics are annotated flow diagrams representing physical actions and their traces, free of the absolute State concept as far as the system is concerned. Quantum theory typically uses classical logic for predicates about the experimenter and apparatus for these are of such low resolution that quantum logical corrections cannot show up.

On the whole it seems harder to accept the relativity of Truth than of Time, and it takes longer. Sometimes to break free of the absolute State it helps to drop the state terminology altogether, and call the actual i/o processes what they are: modes (of action). But if one remembers that states are relative to experimental frames, and there is no state variable of the system any longer, it is comfortable to continue using the familiar word, just as we still speak of time in special relativity.

In quantum physics the classical concept of State variable of the system first splits into the two input and output modalities of determination.

How many experimenters does it take to do a quantum experiment?

Two: one to input the system and one to output it. But one physicist can do both if the experiment is small enough.

Then the two pieces fragment further into the infinitely many mode variables of infinitely many possible experimenters, who may be of either i/o modality. Each experimenter has a family of mutually exclusive, together exhaustive, modes of filtration, or channels, all input or all output. These multiple modal distinctions are ignored in classical logic. There Truth is Truth.

When we speak of position of a particle we could mean either a variable x or any of its values 0, 1 mm, 2 mm, 3 mm, ... Here we sometimes have to say whether we mean variables or their values. One is a process, the other a result of the process.

Each quantum experimenter has a choice among the possible modes of ideal determinations by that experimenter on that system. The possible modes are together exhaustive, mutually exclusive, and each maximally informative or sharp. In classical physics a predicate about the system could be any set of states of the system; in quantum physics one represents an input process by a special kind of set that contains the entire straight line through any two of its points; a linear space. The generic set has no physical interpretation.

Quantum i/o modes correspond to terms in a complete resolution of the unit operator into one-dimensional projection operators (or matrices) P_n obeying the familiar corresponding conditions

$$1 = \sum P_n, \quad P_n^2 = P_n = P_n^\dagger, \quad P_n P_m = 0, \quad n \neq m, \quad \text{Tr } P_n = 1.$$

The already-given logical meanings of these conditions are the same for q logic and c; except for the $P = P^\dagger$ condition, which expresses a temporal symmetry between input and output that is so deeply built into the atemporal c logic that it never needs mention there. The mode variable or relative state variable for an experimenter is the variable whose possible values are the P_n, or the index n alone will do. One can learn the relative state variable by examining the experimental process. One determines a relative state value by doing the experiment.

A physical quantity is an operator in the same linear algebra as the modes P_n.

Quantum theory predicts transitions $\langle b|a\rangle$ and values that observables that can have. The quantum formulae for these predictions, suitably specialized, cover the classical case, but the classical concepts cannot cope with the quantum experience, as is typical for expansive reforms.

The possible values of a variable quantity Q (relative to the mode of determination) are the roots q of the equation

$$QP = qP, \quad P \neq 0. \tag{1}$$

The instantaneous transition probability is

$$\mathrm{Prob}(P \leftarrow Q) = \frac{\mathrm{Tr}\, PQ \,\mathrm{Tr}\, QP}{\mathrm{Tr}\, QQ \,\mathrm{Tr}\, PP}. \tag{2}$$

One may associate a vector $|\psi\rangle$ with the input mode $P = |\psi\rangle\langle\psi|$ and a dual vector $\langle\phi|$ with the output mode $Q = |\phi\rangle\langle\phi|$ and define a transition amplitude

$$A = \langle\phi|\psi\rangle, \quad \mathrm{Prob}(P \leftarrow Q) = |A|^2. \tag{3}$$

In this context, P and Q are not variables of the system nor are $\langle\phi|$ and $|\psi\rangle$. They are different values of different mode variables. They represent actions of the two experimenters, one who inputs systems through a P filter and one who outputs them through a Q filter, counting the systems that they input or output. The probability predicts the ratio of the output counts to the input.

All the equations (1-2) hold in both c and q logic. Where is the difference?

In the c logic all the projections and filters commute, $PQ = QP$, so do all the variables, and one resolution of unity works for all. Experimenters at most permute each others modes. Modes of different i/o modality can be identified in pairs.

In the reformed q logic this is the commutativity that breaks down to a degree measured by $\hbar$. This non-commutativity is easily verified by eye with three polarizers. Two experimenters of the same i/o modality are related by the most general norm-preserving invertible linear transformation. In the generic case different experimenters have different mode variables, different relative state variables. One cannot translate one language into the other. One can only give transition probabilities from the input modes of one language to the output modes of another. This corresponds to the dictionary practice of labeling some meanings as rare, others as common. The mode values of one experimenter always commute with one another as in c logic, but rarely with those of another experimenter.

In a multichannel experiment an experimenter can determine which one and only one of his or her channels passes a system. In a one-channel experiment, the experimenter can determine whether the system passes that channel. The multichannel experiment answers the question , "What is the state relative to my mode of analysis?" The one-channel answers the question "Is the state P_1 relative to my mode of analysis?" The values of the state correspond to input or output filter operations. There is no experiment on a q system to answer a question "What is the state of the system?" as there is for a c system.

Talk of collapse arises when a q input mode is mistaken for a c State, which is always both input and output. Suppose that with 0 time delay the system passes some arbitrary output filtration process. In c logic the two states are then the same State. Imagine for a moment that for a q system the input mode defines a True State. We must then compensate for this error with a second: that the initial "State" "collapses" to the final "State." There is no such collapse in quantum physics because there is no state to collapse, just an input process and an output process. The input process does not run down the optical bench and collapse into the output process. The filters stay put and the photon runs down the optical bench from the input filter to the output. There is no information transmitted down the bench except at the speed of the photon.

It is not that we are denied information of the State, any more than of the Time; rather, there is no physical reason to suppose that one exists, other than our prejudice for the singular.

None of this is new. It is all expressed in the usual quantum theory by saying that the observable variables Q are all the (normal) operators on the system space and the possible values of any Q are the roots of the equation

$$Q|\psi\rangle = q|\psi\rangle, \quad |\psi\rangle \neq 0.$$

This excludes the possibility of some State variable Ψ that assigns a different value to every mode vector $|\psi\rangle$, so that a measurement of Ψ could then be a determination of a value $\psi\rangle$. (For if every vector assigns a value to the operator Ψ then $\Psi \equiv $ const and its measurement determines no state at all.) If observables are non-commuting operators, then "the state vector" is no observable. To be sure, each individual mode vector ψ defines a binary observable $P = \psi \otimes \psi^\dagger$ whose value is 0 or 1. P stores only 1 bit of information, not enough to be a determination of some many-valued hypothetical state variable.

Likewise one usually cannot properly speak of a q variable as simply "having" a value, but usually must distinguish between having an input value or an output value. Otherwise we run into contradictions when a photon flies between obliquely oriented polarizing filters. To be sure, when we speak as if the system had a state, we must speak as if the variables have values. it is conventional to call the input value the value for the whole trip down the optical bench and call the output value the value-after-collapse.

Quantum logic is a temporal logic, a language of verbs for actions, not nouns for states. It is closer to experimental practice to speak of its vectors ψ as modes (of action) or channels, not states (of being). But language has a stability of its own. Over half a millennium since Copernicus we still speak of sunrise but use a fixed sun to plot the planetary orbits. Probably we will similarly always speak of quantum states but use them as modes. But making states collapse when the quantum system passes from an input channel to an output one is letting the word be the master.

3 Logical Algebra

There is no essential difference between logic and set theory [14].

Classical logic or set theory is a jack-of-all-trades, used to make all mathematical models. Heisenberg and Bohr replaced the classical logic as a tool of thought by their non-commutative matrix algebra of observables, which includes non-commuting predications, and is already a working quantum logic. For quantum models we require a quantum logic that is as versatile as classical logic and also incorporates the matrix logic of quantum theory today.

There is a logic meeting these requirements that works for both c and q physics. It corresponds closely enough to classical set theory to be called quantum set theory. It derives from how one constructs aggregates of quanta obeying the exclusion principle. One may call such quantum aggregates quantum sets.

Quantum set theory is based on real Clifford algebra Classical set theory of finite sets can be formulated as a Clifford algebra over the binary field 2. A Clifford algebra is an associative algebra of polynomials, in a finite number of variables called generators forming a vector space, obeying the Clifford law:

It has no useful classical analogue. The operator algebra of a squad is a Clifford algebra.

We suppose that the cosmos is a q set of bits, not a squad of bits. We tabulate some correspondences between the Clifford logic and the Boole logic, two contractions removed:

EXPANSION	$\longrightarrow$	CONTRACTION
Clifford algebra	$\longrightarrow$	Boolean algebra
q set	$\longrightarrow$	c set
γ's	$\longrightarrow$	bits
$+$	$\longrightarrow$	(nil)
Clifford product $\sqcup$	$\longrightarrow$	XOR
Grassmann product $\vee$	$\longrightarrow$	POR
Grade	$\longrightarrow$	Cardinality

4 Higher-Order Logic

To define sets of sets one merely iterates. If C is a Clifford algebra, taken as a first-order logic of sets, then the second-order logic of sets of sets is represented by the second Clifford algebra $\mathbf{2}^C$. We use the quadratic form $\|x\| := \operatorname{Re} x^2$ on C to define the Clifford algebra over C. Starting from $C^0 := \mathbb{R}$, we generate the infinite sequence of the natural algebras

$$\mathbf{C}^{n+1} = \mathbf{2}^{\check{C}^n} \tag{5}$$

$$\begin{aligned}
\mathbf{C}^0 &= \operatorname{Cliff}(0,0) = \mathbb{R}, \\
\mathbf{C}^1 &= \operatorname{Cliff}(1,0) = 2\mathbb{R}, \\
\mathbf{C}^2 &= \operatorname{Cliff}(2,0) = \overline{\mathbb{Q}}, \\
\mathbf{C}^3 &= \operatorname{Cliff}(3,1) = \mathbb{D}, \\
\mathbf{C}^4 &= \operatorname{Cliff}(10,6), \\
\mathbf{C}^5 &= \operatorname{Cliff}(2^{15}+2^7, 2^{15}-2^7), \\
&\vdots \\
\mathbf{C}^\infty &= \lim_{n\to\infty} \mathbf{C}^n = \bigcup_n \mathbf{C}^n
\end{aligned}$$

For all the natural algebras

$$(\text{Signature})^2 = \text{Dimension}.$$

For all causal spaces

$$|\text{Signature}| = \text{Dimension} - 2.$$

So the only causal natural algebra is the 4-dimensional $\mathbf{C}^2$.

The limit $\mathbf{C}^\infty$ has physical meaning only as a set of theoretical possibilities. It is the free Clifford algebra generated by $\mathbb{R}$ and a free operator $\iota : \mathbf{C}^\infty \to \mathbf{C}^\infty$ that maps any non-zero element 0 into a generator of the same norm. The operator *iota* is called the unifier, since it maps the product of any number of sets into a unit set. It is the q descendant of the successor operator ι that Peano used to define the natural numbers; hence the term "natural" Clifford algebra.

5 Stable Calculus

The algebra of the differential calculus is based on the relations

$$\begin{aligned}{}[\partial, x] &= I \\ [I, \partial] &= 0 \\ [x, I] &= 0 \end{aligned} \tag{6}$$

These define a three-dimensional real Lie group that can be called the Newton group; its complexification is the Heisenberg group $\mathbf{H}(3)$ of complex 3×3 matrices of the form

$$\begin{bmatrix} 1 & a_{12} & a_{13} \\ 0 & 1 & a_{23} \\ 0 & 0 & 1 \end{bmatrix}$$

It is unstable. The source of this Newton instability is plain. The point $p = (x, y)$ is simple and stable. The chord $(p, p + \Delta p)$ is semisimple and stable. The limiting tangent vector (p, dp) is compound and unstable. The infinitesimal limit $\Delta x \to 0$ is what destabilizes.

The regular relations, suitably rescaled, are

$$\begin{aligned}{}[\partial, x] &= I \\ [x, I] &= \pm \partial \\ [I, \partial] &= x \end{aligned} \tag{7}$$

They define either $SO(3)$ (for the positive sign) or $SO(2, 1)$ (for the negative sign). The $SO(2, 1)$ Clifford algebra is the algebra of 2×2 real matrices. The $SO(3)$ Clifford algebra is the algebra of the quaternions (1×1 quaternionic matrices). We take the $SO(2, 1)$ case for illustration. We can satisfy these commutation relations for $SO(2, 1)$ with the real Dirac matrices $x \sim \gamma_{14}$, $\partial \sim \gamma_{12}$, $I \sim \gamma_{24}$. This is the smallest free Clifford algebra that will do.

The spectra of these 4×4 matrices are too small to approximate the differential operators x, ∂, but we can enlarge the spectrum by using the direct sum of many replicas. This gives us an algebraic substitute for the differential calculus that approximates it as closely as we like and has an exact symmetry group as close to the Newton group as we like.

The Heisenberg instability permeates canonical quantum mechanics, boson theory, gauge theory, and general relativity. One stabilizes all these instabilities at once by reforming the Heisenberg group to satisfy a single principle essentially implied by Segal:

6 Simplicity Principle

The transformation group of a physical theory is a simple Lie group.

The transformation group of classical mechanics is the canonical group, which is not even a Lie group, since its elements depend on arbitrary functions.

The transformation group of standard quantum theory is the unitary group of its Hilbert space. This is a simple Lie group just when the space if finite-dimensional: for example, for spins, polarizations, spin networks, quantum computers, but not for a linear harmonic oscillator. The time-independent linear harmonic oscillator has been stabilized by group regularization so that the result obeys the simplicity principle [1, 2, 13, 5, 20]

We leave it as an exercise to develop the first-order quantum logic from the simplicity principle. The operator algebra of the quantum theory is presumably the embedding algebra of the Lie algebra of the simple group. The centrality of i makes the unitary and symplectic groups unstable, leaving only the orthogonal groups for systems with Lie groups of high dimensionality.

Let us list the Lie groups of the standard model and gravity for instability, marking the regular ones • and the singular ∘.

- SO(2) of complex amplitudes
- SO(2) of hypercharge
- SO(3) of spin
- SU(3) of color

- SO(3,1) of spin
- ø Gauge groups of the standard model
- Fermion field operators
- ø Boson field operators
- ø Space-time diffeomorphisms (gravity).

This is also a to-do list. We must stabilize the unstable groups and fill in the open circles.

Our recent work on group regularization has centered on time-dependent quantum theory, where the centrality of time t is unstable.

We have stabilized the Heisenberg algebras of space and time of the linear harmonic oscillator and the Dirac spin-1.2 particle.

To regularize a field theory we must construct quantum functions of quantum points. This already comes up in the oscillator, whose quantum coordinate q is a function of quantum time t, and $q(t)$ is easily constructed [19]

Clifford algebra provides a general practical solution to this problem. The time-dependent coordinates of c theory belong to the space S^T where S is the c space axis and T is the c time axis. The problem is to define S^T when both are quantum. In that case we define the spaces by their simple algebras. What is the simple algebra S^T when S and T are simple algebras?

The difficulty evaporates if S is a Clifford algebra $S = \mathbf{2}^{\triangle S}$ of a generator space $\triangle S$. Then we set

$$S^T = (\mathbf{2}^{\triangle S})^T := \mathbf{2}^{\triangle ST}. \tag{8}$$

This is not simple, due to the presence of two independent algebras $\triangle ST$. To simplify it we suppose that these two spaces are also Cliffford algebras,

$$\triangle S = \mathbf{2}^{\triangle^2 S}, \quad T = \mathbf{2}^{\triangle T} \tag{9}$$

Then

$$S^T = (\mathbf{2}^{\triangle S})^T := \mathbf{2}^{\triangle ST} = \mathbf{2}^{\tilde{2}(\triangle^2 S \oplus \triangle T)}. \tag{10}$$

This is still compound, but in a familiar way. It looks as if a simple space $\triangle^2 U = \triangle^2 S \oplus \triangle T$ has spontaneously reduced its symmetry to a direct sum, as by a condensation process.

Our assumption that the i/o processes of the cosmos are generators of a cosmic Clifford algebra U is an atomic hypothesis:

7 Atomic Hypothesis

History is composed of elementary or atomic operations — operons — of unit norm.

The generators γ are the operons (elementary q events, q bits) of history. The bits of classical computers are distinguished by their addresses. The cosmic q bits are distinguished by their internal structure, which can be regarded as a kind of address, describing how one gets to the bit from the origin 1 by a succession of the basic steps of union (XOR) and unification ι).

Now the transformation group of the cosmos is an orthogonal group $SO(N_+, N_-)$, which we suppose has signature $N_+ - N_- \ll N_+ + N_-$, so that the bits of the cosmos can still have nearly Fermi-Dirac statistics. This regular quantum theory introduces two regularization parameters, which can be taken to be the number N of γ's in the cosmos and the duration χ of one operon γ.

We regularize the c space-time operators x^μ, $p_\mu = -i\hbar\partial_\mu$ of a Dirac quantum into the finite quantum operators

$$\widehat{x}^\mu = i\chi \sum_{n=1}^{N} \gamma^{\mu 5}(n),$$
$$\widehat{p}_\mu = -i\epsilon \sum_{n=1}^{N} \gamma^{\mu 6}(n),$$
$$\widehat{r} = N^{-1} \sum_{n=1}^{N} \gamma^{56}(n)$$

with

$$\hbar = \frac{N^2 \chi \epsilon}{2}, \quad \hbar' = \frac{\chi}{N\epsilon}, \quad \hbar' = \frac{\epsilon}{N\chi},$$

We diagram the Clifford algebra generators that we used for the Dirac particle with the following "chronosome." We have arbitrarily restricted ourselves to the natural algebras. This diagram is only one of many possible structures whose distinguishing physical consequences are still unknown.

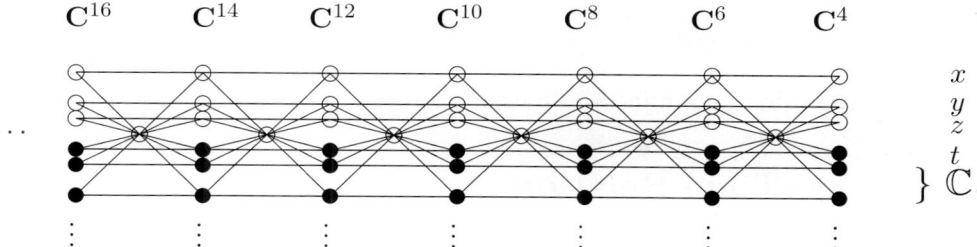

KEY
- ○ spacelike operons γ
- ● timelike operons γ
- — causal links ι (right to left)
- ⋮ 10 unused γ's (GUT?)

Each basic Clifford element has such a diagram, but their superpositions soon make the diagrams unwieldy.

It is possible to compare these Clifford models with supersymmetric ones, at least in their most salient aspects. Supersymmetric variables consist of body and soul, in the terminology of Bryce DeWitt. The body is bosonic, the soul fermionic. The regular quantum theory instead analyzes all body into soul, by "bosonizing" an aggregate of many fermionic spins. Its quantum variables are all soul.

Group regularization, like quantization and supersymmetrization, is a process that we apply to a working theory. It changes the algebraic relations slightly, thereby stabilizing them. Supersymmetrization preserves all the old instabilities intact and introduces some new particles, a super-partner for every particle. Group regularization seems to introduce one particle, the quantum of the $\hat{r}$ variables, which is presumably the Higgs particle.

As an initial crude bound on the regularization constants, we can say that $N 10^{13}$, the photon count in one mode of a laser, and $\epsilon \sim 10^2$ GeV, if we take the top quark mass as the top of the particle mass spectrum.

In general
$$\Delta \hat{x} \Delta \hat{p} \geq |\hat{r}| \hbar / 2$$
In the contraction limit $|\hat{r}| \to 1$,
$$\Delta \hat{x} \Delta \hat{p} \geq \hbar / 2$$

In $\hat{r}$ melt-down, $|8r| < 1$ and
$$\Delta\hat{x}\Delta\hat{p} < \hbar/2$$
can occur. Variable $\hat{r}$ manifests in part as reduced effective $\hbar$.

8 Energy-Time Relation

Each of the reforms listed in section 1 introduces a form of energy whose scale is set by the regularization constant. For example, the k regularization brings in the energy $E = kT$ of an ideal linear harmonic oscillator at temperature T, and the most famous equation of physics gives the energy $E = mc^2$ associated with a rest mass m by the c regularization. What is the new energy of the finite quantum reform?

The regular group $SO(10, 4)$ transforms time changes to energy changes in the relation
$$\Delta E \sim \Pi \Delta t, \quad \Pi = \frac{\epsilon}{\chi},$$
with fundamental power Π determined by the ratio of the ergon ϵ to the chronon χ. On dimensional grounds Π should be on the order of the Planck power, 10^51 W. This group symmetry is broken by the usual vacuum but restored by space-time melt-down. While this condition seems too extreme for practical application, the process might occur in astrophysical situations like the creation of the universe or the core of a black hole.

Acknowledgments

For Mohamed El Naschie. Based on a presentation at ZKM Karlsruhe, Oct. 11, 2003. This presentation is based on work done with James Baugh, Andrei Galiautdinov, Heinrich Saller, and Mohsen Shiri-Garakani. It derives from seminal ideas of John Von Neumann, David Bohm, Roger Penrose, Richard Feynman, and I. E. Segal.

References

[1] Arecchi, F. T., E. Courtens R. Gilmore, and H. Thomas, *Physical Review* **A6**, 2211 (1972).

[2] Atakishiyev, N. M., G. S. Pogosyan, and K. B. Wolf. Contraction of the finite one-dimensional oscillator. *International Journal of Modern Physics* **A18**, 317-327 (2003).

[3] J. Baugh, D. R. Finkelstein, A. Galiautdinov, and M. Shiri-Garakani. Transquantum dynamics. *Foundations of Physics* **33** 1267 (2003)

[4] Carlen, E. A. and E. H. Lieb. Optimal hypercontractivity for Fermi fields and related non-commutative integration inequalities. *Communications in Mathematical Physics* **155** 27 (1993)

[5] Carlen, E. and R. Vilela Mendes. Non-commutative space- time and the uncertainty principle. arXiv:quant-ph/0106069 v1 12 Jun 2001 *Physics Letters* **A290**, 109 (2001)

[6] Finkelstein, D., J.M. Jauch, S. Schiminovich and D. Speiser. Some physical consequences of general Q-covariance. *Helvetica Physica Acta* 35, 328-329 (1962)

[7] Finkelstein, D., , J.M. Jauch, S. Schiminovich and

D. Speiser, Principle of general Q-covariance, *Journal of Mathematical Physics* **4**, 788- 796 (1963)

[8] D. Finkelstein, *Quantum Relativity*. Springer-Verlag, Heidelberg (1996)

[9] Finkelstein, D., A. Galiautdinov. Cliffordons. *Journal of Mathematical Physics* **42**, 3299-3314 (2001).

[10] Finkelstein, D.R. What is a photon? *Optics and Photonic News*, in press (2003).

[11] Galiautdinov A.A., and D. R. Finkelstein. Chronon corrections to the Dirac equation. hep-th/0106273 [LANL]. *Journal of Mathematical Physics* **43**, 4741 (2002)

[12] Inönü, E. and E. P. Wigner. On the contraction of groups and their representations. *Proceedings of the National Academy of Sciences* **39**(1952) 510-525.

[13] Kuzmich, A., N. P. Bigelow, and L. Mandel, *Europhysics Letters* **A 42**, 481 (1998). Kuzmich, A., L. Mandel, J. Janis, Y. E. Young, R. Ejnisman, and N. P. Bigelow, *Physical Review A* **60**, 2346 (1999). Kuzmich, A., L. Mandel, and N. P. Bigelow, *Physical Review Letters* **85**, 1594 (2000).

[14] Morse, A.P. *A Theory of Sets*. Academic Press (1986)

[15] Tavel, M, D. Finkelstein, and S. Schiminovich, Weak and electromagnetic interactions in quaternion quantum mechanics, *Bulletin of the American Physical Society* **9**, 436 (1965)

[16] 'tHooft, G. Determinism in Free Bosons. *International Journal of Theoretical Physics* **42**, 355 (2003)

[17] Segal, I. E. . A class of operator algebras which are determined by groups. *Duke Mathematics Journal* **18** (1951) 221.

[18] Segal, I.E.: A non-commutative extension of abstract integration, Annals of Math., **57** (1953) 401–457.

[19] Shiri-Garakani, M. and D. Finkelstein. Finite quantum harmonic oscillator. Submitted 2004.

[20] Vilela Mendes, R. "Quantum mechanics and non-commutative space-time." *Physics Letters* **A210**, 232 (1994). *Journal Of Physics A: Math, Gen*, **27**, 8091 (1994). *Physics Letters* **A210**, 232 (1996). *J. Math Phys.* **41**, 156 (2000).

Atomic Cluster Physics: New Challenges for Theory and Experiment

by Walter Greiner and Andrey Solov'yov

Abstract

A brief introduction to atomic cluster physics, the interdisciplinary field, which developed fairly successfully during last years, is presented. A review of recent achievements in the detailed *ab initio* description of structure and properties of atomic clusters and complex molecules is given. The main trends of development in the field are discussed and some of its new focuses are outlined. Particular attention is devoted to the role of quantum and many-body phenomena in the formation of complex multi-atomic systems and the methods of theoretical investigation of their specific properties. The role of the simplified model approaches accurately developed from the fundamental physical principles is stressed. Various illustrations are made for sodium, magnesium clusters, fullerenes and clusters of noble gas atoms.

1 Introduction

During the last decade it was recognized, both experimentally and theoretically, that complex molecules and atomic clusters (ACs) often possess unique properties, which make them a new object of physical research, rather different from both a single atom and from the solid state. The knowledge of the detailed electronic and ionic structure of single complex molecules and nano-clusters can be essential for various practical applications, such as the formation of new materials, nano-structures, in the design of drugs and biologically active spices as well as for the understanding the fundamental issues, such as functioning quantum and thermodynamic laws in the nano-scale systems or mechanisms for complex multi-atomic systems formation, self-assembly and functioning.

The demand in the understanding of principles of assembly and functioning of complex multi-atomic systems such as bio-molecules or nano-clusters is tremendous, because of the potential use of this knowledge for purposes of microelectronics, of biochemistry, the drug industry etc. The problems of self-organization, of self-assembly and of the functioning of complex multi-atomic aggregates and their interactions have been addressed both theoretically and experimentally in a large number of papers from different perspectives. Often, these problems can be reduced to the problem of the interaction of a limited number of atoms within a complex molecule or even to the interaction of a single atom or ion with a certain fragment of a complex molecule (an active center responsible for a certain function) or a cluster structure. Thus, in order to achieve a real breakthrough in the field, one needs to learn how to handle (both theoretically and experimentally), i.e. to be able to manipulate experimentally and predict theoretically properties of multi-atomic systems containing about 100 atoms. or maybe a little less or a little more than this limit. With this knowledge in hand, one can then move towards a detailed *ab initio* understanding of the properties of larger multi-atomic systems, biomolecules (proteins, DNA), which typically consist of rather small fragments (amino acids or bases) whose structure and interactions do not involve more than 100 atoms.

These structures are, nowadays, subjects of very intensive theoretical and experimental studies in many physical, chemical and biological laboratories and institutions worldwide. A variety of methods have been used to investigate these objects (see, e.g., [1, 2, 3, 4, 5, 6, 7, 8, 9, 10] and references therein). Due to these efforts a vast amount of physical, chemical and biological data on the properties of complex multi-atomic systems and their interaction with the environment have been accumulated. However, it can be stated that until now there is no consistent theoretical approach, based solely on the fundamental principles of quantum physics, which might allow one, not only to explain systematically the known experimental data, but also, and this is quite essential, to predict new properties of the objects and new phenomena related to them. Nearly all theoretical approaches developed so far can be termed 'phenomenological' in the sense that each one of them substitutes the full quantum-mechanical description of the dynamics of constituents of a multi-atomic structure with a model theory which uses a set of parameters deduced from the experimental data. Each of these models is able to reproduce a limited number of particular properties of a complex multi-atomic system of a particular type, since the sets of the parameters involved are not of a universal nature. Thus, the model theories have severe restrictions: in each case they can explain but few of the experimental data. From them, often, one can hardly draw general conclusions or produce predictions on the properties of other structures. A more accurate description of the electronic and ionic structures, internal dynamics and interaction with external objects and fields has to be elaborated. The development of such an approach, which is multifaceted and includes not only theoretical investigations based on the first principles of the quantum many-body theory but also implies a great amount of experimental work and computing, is a subject of current joint efforts of specialists in various fields of physics and chemistry.

The complete theoretical description of nano-scale systems consisting of about 100 atoms is extremely difficult [5]. So far, an *ab initio* many-body quantum mechanical description accounting for all electrons in the system can be used effectively for systems of a few tens of atoms [6, 7] rather than hundreds. The computer power required for such calculations grows exponentially with increasing molecular or cluster size. Therefore, one needs to invoke various simplified model approaches in order to describe complex systems of sufficiently large size [5, 8, 9].

In this paper, the role of the careful choice of the model and the importance of accounting for many-body and quantum phenomena is demonstrated for the process of formation of the AC magic numbers within the cluster size range $N < 150$. Here N is the number of atoms in the system. This example shows that the high predictive power of a model can be achieved on the basis of detailed comparison of the predictions of the model and *ab initio* approaches with each other and with experiment for relatively small systems, consisting of tens of atoms, and by the extrapolation of the model postulates towards larger scale systems [5, 6, 7].

In order to illustrate these ideas, in the next sections, we briefly discuss fission, fusion and collision processes involving ACs as well as some general aspects of AC science.

2 Atomic cluster science

A group of atoms bound together by interatomic forces is called an atomic cluster. There is no qualitative distinction between small clusters and molecules. However, as the number of atoms in the system increases, ACs acquire more and more specific properties making them unique physical objects different from both single molecules and from the solid state.

In nature, there are many different types of AC: van der Waals clusters, metallic clusters, fullerenes, molecular, semiconductor, mixed clusters, and their shapes can depart considerably from the common spherical form: arborescent, linear, spirals, etc. Usually, one can distinguish between different types of clusters by the nature of the forces between the atoms, or by the principles of spatial organization within the clusters. Clusters can exist in all forms of matter: solid state, liquid, gases and plasmas.

In figure 1, we present images of a few clusters in order to show a big variety of the cluster forms existing in nature. We also show the structure of the $\alpha/\beta-$ triosophosphate-isomerase globule aiming to stress that complex molecules such as proteins can be treated as clusters of subunits and that each

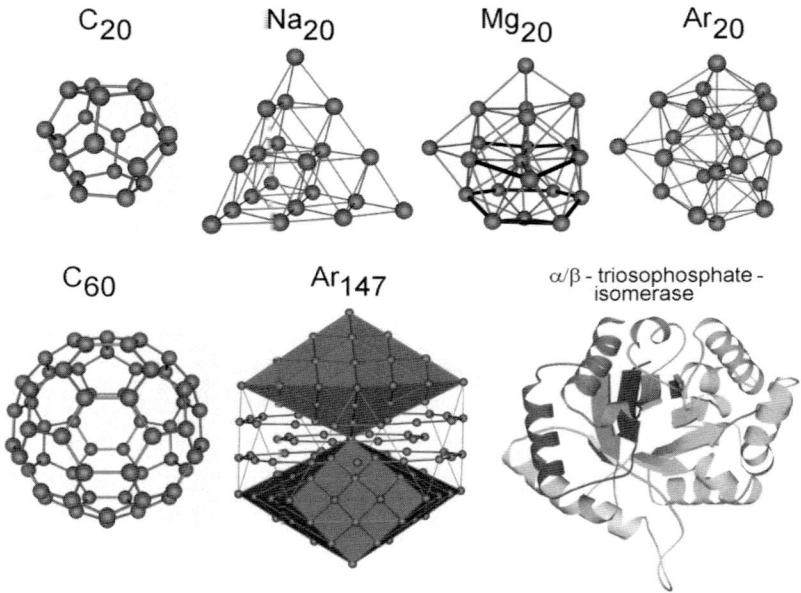

Figure 1: Different nature of forces between the atoms results in different principles of their organization within clusters and complex molecules. Geometries of the presented ACs have been calculated in [6, 7, 5, 10], structure of the protein globule ($\alpha/\beta-$ triosophosphate-isomerase) is taken from [4].

of the subunits is a cluster on its own.

The novelty of AC physics arises mostly from the fact that cluster properties explain the transition from single atoms or molecules to the solid state. Modern experimental techniques have made it possible to study this transition. By increasing the cluster size, one can observe the emergence of the physical features in the system, such as plasmon excitations, electron conduction band formation, superconductivity and superfluidity, phase transitions, fission and many more. Most of these many-body phenomena exist in solid state but are absent for single atoms.

The science of clusters is a highly interdisciplinary field. ACs concern astrophysicists, atomic and molecular physicists, chemists, molecular biologists, solid-state physicists, nuclear physicists, plasma physicists, technologists all of whom see them as a branch of their subjects but cluster physics is a new subject in its own right.

Significant progress achieved in the field over the past two decades brought the understanding of ACs as new physical objects with their own distinctive properties. This became clear after such experimental successes as the discovery of the fullerene C_{60}, of the electronic shell structure in metal clusters, the observation of plasmon resonances in metal clusters and fullerenes, the observation of magic numbers for various other types of clusters, the formation of singly and doubly charged negative cluster ions and many more. A complete review of this field can be found in review papers and books, see e.g. [1, 2, 3, 11, 12, 13, 14, 15, 16].

3 Distinctive properties of atomic clusters: cluster magic numbers

ACs, as new physical objects, possess some properties, which are distinctive characteristics of these systems. The cluster geometry turns out to be an important feature of clusters, influencing their stability and vice-versa. The determination of the most stable cluster forms is not a trivial task and the solution of this problem is different for various types of cluster. The stability of clusters and the their transformations is a theme which does not exist at the atomic level and is not of great significance for solid state but is of crucial importance for AC systems. This problem is closely connected to the problem of cluster magic numbers.

The sequence of cluster magic numbers carries essential information about a cluster's electronic and ionic structure. Understanding the magic numbers of a cluster is pretty well equivalent to understanding its electronic and ionic structure [5]. A good example of this kind occurs for sodium clusters. In this case, the magic numbers arise from the formation of closed shells of delocalised electrons, one from each atom (see [11, 15] and references therein). Another example is the discovery of fullerenes, and in particular the C_{60} molecule [17], by means mass spectroscopy of carbon clusters.

In figure 2, we present the mass spectra measured for Ar and Na clusters (see [11, 13] and references therein), which clearly demonstrate the emergence of magic numbers. The forces binding atoms in these two different types of clusters are different. The argon (noble gas) clusters are formed by van der Waals forces, while atoms in the sodium (alkali) clusters are bound by the delocalized valence electrons moving in the entire cluster volume. The differences in the inter-atomic potentials and pairing forces lead to the significant differences in structure between Na and Ar clusters, their mass spectra and

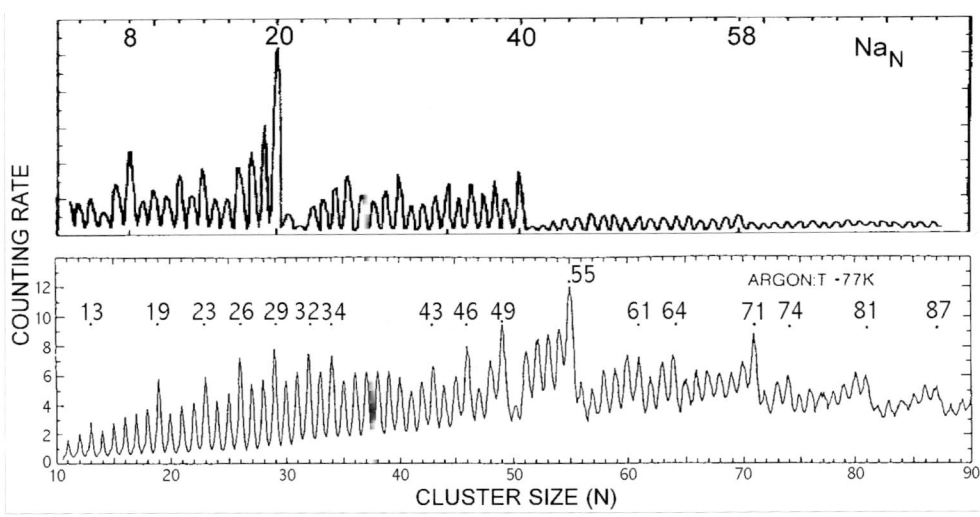

Figure 2: Mass spectra measured for Ar and Na clusters (see [11, 13, 16] and references therein). The intense peaks indicate enhanced stability.

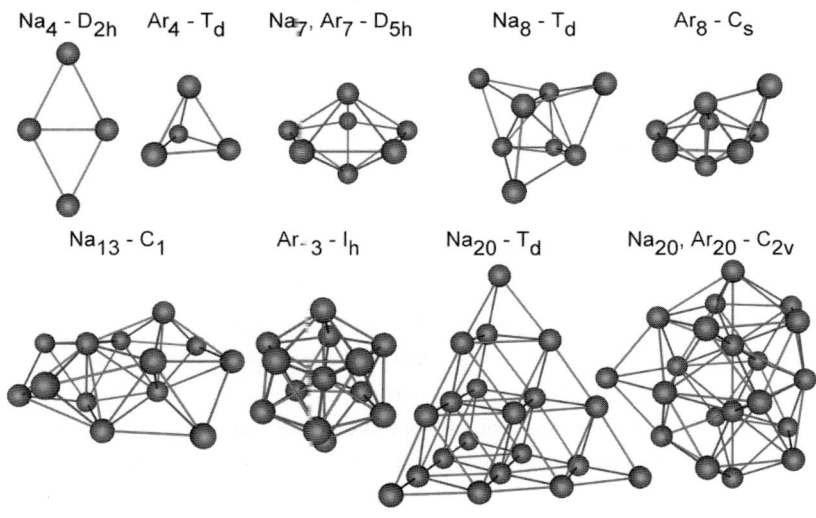

Figure 3: Geometries and the point symmetry groups of some Na and Ar clusters calculated in [5, 6].

their magic numbers.

In figure 3, we present and compare the geometries of a few small Na and Ar clusters of the same size. It is clear from figure 3 that different principles of cluster organization result in different geometries of the alkali and noble gas cluster families.

Such differences can easily be explained. The van der Waals forces lead to enhanced stability of cluster geometries based on the most dense icosahedral packing. The most prominent peaks in mass spectra of argon clusters correspond to completed icosahedral shells of 13, 55, 147, 309 ets atoms. The origin of these magic numbers can be understood on the basis of the classical equations. The origin of the sodium cluster magic numbers is different and is based on the principles of quantum mechanics. In this case the cluster magic numbers 8, 20, 34, 40, 58, 92 ets correspond to the completed shells of the delocalised electrons: $1s^2 1p^6 1d^{10} 2s^2 1f^{14} 2p^6$ ets. This feature of small metal clusters make them qualitatively similar to atomic nuclei for which quantum shell effects play the crucial role in determining their properties [18].

The enhanced stability of cluster systems can be characterized by computing the second differences in cluster binding energies. In figure 4 we present Ar clusters binding energies and their second differences calculated in [5]. The correspondence of the peaks in figure 4 to those in the Ar clusters mass spectrum shown in figure 2 is readily established.

Finally, let us stress the obvious connection between AC physics and physics and chemistry of large molecules, such as proteins or DNA, which in fact can be treated as large clusters of amino acids or bases. The characteristic size of a fragment (amino acid or base) in such clusters is of the order of a few tens of atoms, i.e. the size of a small cluster. It is obvious that the knowledge gained from the AC studies is relevant for the bio-molecular investigations and vice versa. A bunch of interesting phenomena can arise at the juncture of the two fields. For example, fusion of ACs with bio-molecules can create new objects which can be handled as easy as ACs or possess some specific properties and characteristics of ACs, but at the same time carry all essential features of bio-molecules and participate in bio-processes.

4 Collisions involving atomic clusters

Properties of clusters can be studied by means of photon, electron and ion scattering (see Course 9 by A.V Solov'yov in [2, 3] and references therein).

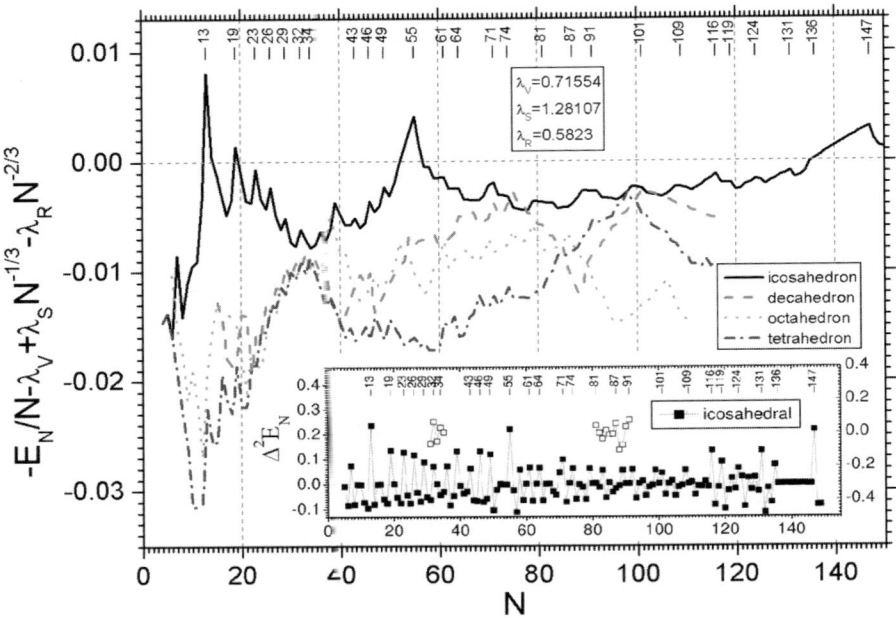

Figure 4: Binding energies and their second differences for Ar clusters calculated in [5].

These methods are the traditional tools for probing properties and internal structure of various physical objects.

Interesting phenomena arise in elastic collisions of electrons with ACs. For example, the diffraction of fast electrons of the fullerene C_{60} molecule was predicted and later observed [19]. The diffraction pattern in the electron elastic scattering cross section caries important information on the electron density in the vicinity of the fullerene's surface.

Electron excitations in metal cluster systems have a profoundly collective nature (see [12] and references therein). They can be pictured as oscillations of electron density against ions, the so-called plasmon oscillations. This name is carried over from solid state physics where a similar phenomenon occurs. Collective electron excitations have also been studied for single atoms and molecules. In this case the effect is known under the name of the shape or the giant resonance. The name giant resonance came to atomic physics from nuclear physics, where the collective oscillations of neutrons against protons have been investigated [18].

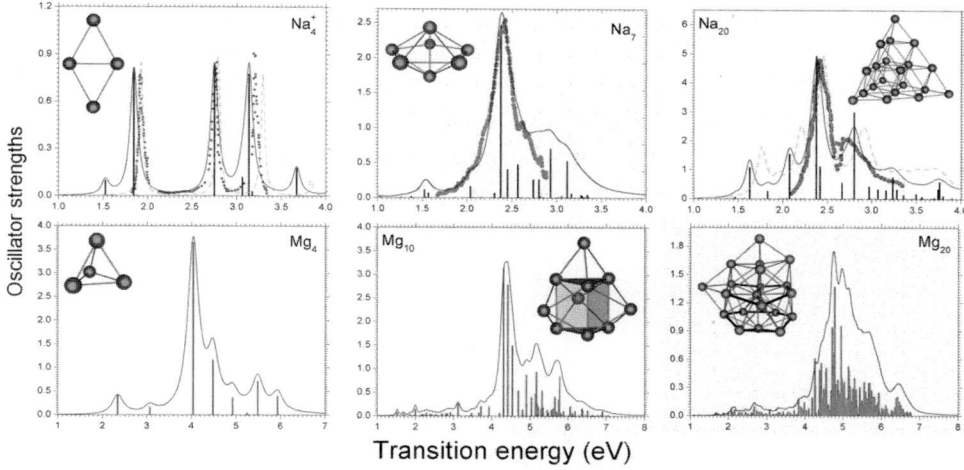

Figure 5: Photoabsortion spectra of some Na and Mg clusters [21].

The interest of plasmon excitations in small metal clusters is connected with the fact that the plasmon resonances carry a lot of useful information about cluster electronic and ionic structure. By observing plasmon excitations in clusters one can study, for example, the transition from the pure classical Mie picture of the plasmon oscillations to its quantum limit or to detect cluster deformations by the value of splitting of the plasmon resonance frequencies.

The plasmon resonances can be seen in the cross sections of various collision processes: photabsorption and photoionization, electron inelastic scattering, electron attachment, bremsstrahlung (see Course 9 by A.V Solov'yov in [2]). Both surface and volume plasmons can be excited. In electron collisions and in the multiphoton absorption regime, plasmons with large angular momenta play an important role in the formation the cross sections of these processes [20].

In figure 5, we present experimentally measured and theoretically calculated cross section for the photoabsoption of some Na and Mg clusters [21]. The cross sections are resonantly enhanced owing to the excitation of plasmon oscillations in the target cluster.

Plasmon excitations in clusters decay via the Landau damping mechanism, while the relaxation of single electron excitations in clusters occurs via the in-

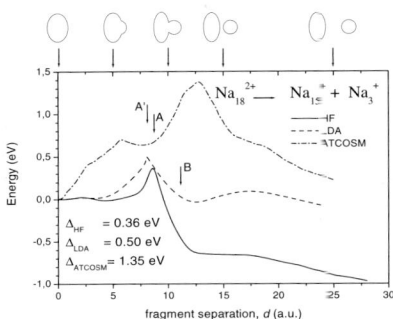 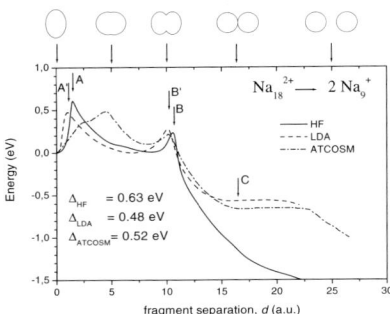

Figure 6: Fission barriers for the asymmetric and symmetric fission channels of Na_{18}^{2+}: $Na_{18}^{2+} \rightarrow Na_{15}^{+} + Na_{3}^{+}$ and $Na_{18}^{2+} \rightarrow 2Na_{9}^{+}$ calculated in [8]

teraction with the vibrations of ions, i.e. via the electron-phonon interaction (see Course 9 by A.V Solov'yov in [2]).

Collisions involving ACs rise many more interesting physical problem. For example, in collisions one can study phase transitions (solid-liquid or liquid-gas) in mesoscopic systems or the cluster multifragmentation process.

5 Fission instability of multiply charged clusters.

Multicharged ACs become unstable towards fission. The process of multicharged metal clusters fission is qualitatively analogous to nuclear fission. The fission instability of charged liquid droplets was first described by Lord Rayleigh in 1882 [22] within the framework of classical electrodynamics. The review of recent work on metallic cluster fission, one can find in [2, 8, 14, 15].

The fission process of ACs is interesting because it reveals the obvious parallel of AC studies with nuclear physics, where the fission process of nuclei has been studied for many decades [18]. The experiments on cluster fission provide a very good opportunity to test various concepts, approximations and AC models. It convincingly demonstrates the importance of the correct accounting for the quantum and many body phenomena in the description of multi-atomic systems. Dynamic aspects of the AC fission problem are also of great interest, because in the contrary to nuclear physics in the fission of ACs all the forces in the system are known and thus one can develop the full dynamic description of the process.

To illustrate the fission of charged metal clusters we plot in figure 5 the fission barriers for the symmetric and asymmetric fission channels of Na_{18}^{2+}: $Na_{18}^{2+} \rightarrow Na_{15}^+ + Na_3^+$ and $Na_{18}^{2+} \rightarrow 2Na_9^+$. The bariers plotted in figure 5 have been calculated in [8] within the two-center LDA and Hartree-Fock jellium model and compared with the asymmetric two-center-oscillator shell model (ATCOSM). Figure 5 demonstrates the evolution of cluster shape during the fission process, the importance of cluster deformations, many-electron correlation and shell effects.

6 Fusion process of atomic clusters.

The formation of a sequence of cluster magic numbers should be closely connected to the mechanisms of cluster formation and growth. It is natural to expect that one can explain the magic numbers sequence and find the most stable cluster isomers by modelling mechanisms of cluster assembly and growth, i.e. the fusion process of ACs [5].

The problem of magic clusters is closely connected to the problem of searching for global minima on the cluster multidimensional potential energy surface. The number of local minima on the potential energy surface increases exponentially with the growth cluster size and is estimated to be of the order of 10^{43} for $N = 100$ [2, 5]. Thus, searching for global minima becomes increasingly difficult problem for large clusters. There are different algorithms and methods of the global minimisation, which have been employed for the global minimisation of AC systems (see [2, 5] and references therein). These techniques are often based on the Monte-Carlo simulations.

Alternatively, the algorithm based on the dynamic searching for the most stable isomers in the cluster fusion process has been recently proposed [5]. The calculations performed with this new algorithm demonstrated that this approach is an efficient alternative to the known techniques of the cluster global minimisation. The big advantage of the fusion approach consists in the fact that it allows to study not just the optimized cluster geometries, but also their formation mechanisms.

In the recent work [5], the fusion algorithm was formulated in a most simple, but general form. In the most simple scenario, it was assumed that atoms in a cluster are bound by Lennard-Jones potentials and the cluster fusion takes place atom by atom. In this process, new atoms are placed on the cluster sur-

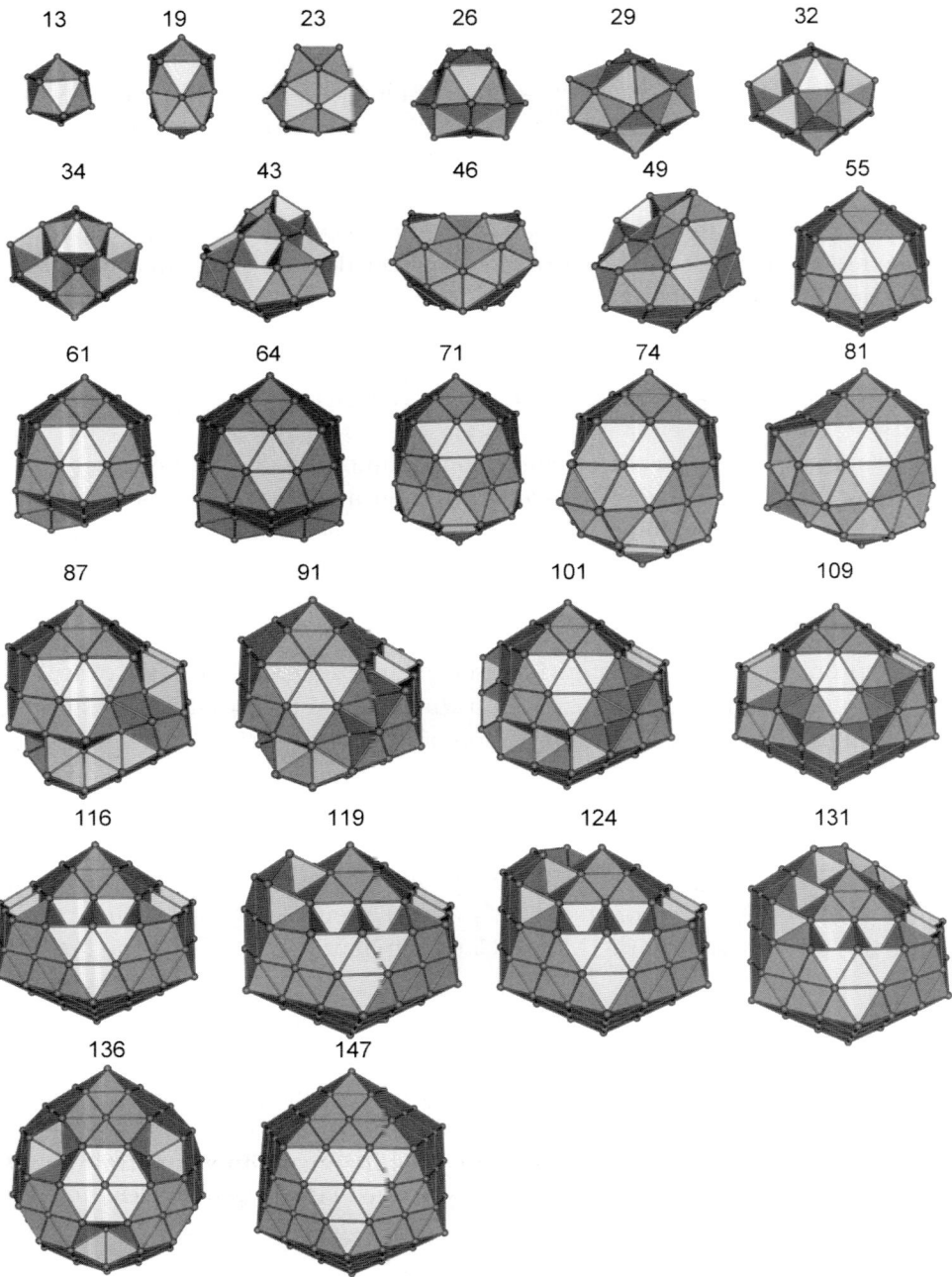

Figure 7: Images of the Lennard-Jonnes global energy minimum cluster isomers [5, 10]. The mass numbers of the pictured clusters correspond to the magic numbers of the noble gas (Ar, Kr, Xe) clusters.

face in the middle of the cluster faces. Then, all atoms in the system are allowed to move, while the energy of the system is decreased. The motion of the atoms is stopped, when the energy minimum is reached. The geometries and energies of all cluster isomers found in this way are stored and analysed. The most stable cluster configuration (cluster isomer) is then used as a starting configuration for the next step of the cluster growing process.

Starting from the initial tetrahedral cluster configuration and using the strategy described above, the cluster fusion paths have been analysed up to the cluster sizes of more than 150 atoms. We have found that in this way practically all known global minimum structures of the Lennard-Jonnes clusters can be determined. Figure 7 shows the images of the Lennard-Jonnes global energy minimum cluster isomers [5]. The mass numbers of the represented clusters correspond to the magic numbers of the noble gas (Ar, Kr, Xe) clusters.

So far, the cluster fusion algorithm has been applied to the noble gas clusters which are based on the LJ type of the inter-atomic interaction. However, the fusion process can be generated in a similar way for the systems, like metal clusters, holding together by quantum forces. This technique can also be used for the simulation of the fusion process of complex bio-molecules (proteins and DNA) or for the study of proteins folding. It would be interesting to see to which extent the parameters of inter-atomic interaction can influence the cluster fusion process and the corresponding sequence of magic numbers or whether the crystallization in the nuclear matter consisting of alpha particles and/or nucleons is possible. Studying cluster thermodynamic characteristics with the use of the developed technique is another interesting theme which is left open for future considerations.

7 Conclusions

In recent years, AC physics has made very significant progress, but a large number of problems in the field are still open. The transition of matter from the atomic to the solid state implies changes of organization which turn out to be a good deal more subtle and complex than was originally supposed. Different type of clusters, composite clusters, various size ranges, cluster geometries, complex molecules (including biological), clusters on a surface and in plasmas, all provide additional themes which make this field of science very rich and varied. Collisions involving ACs, mass spectroscopy and laser techniques provide tools for experimental studies of the cluster structure and properties.

However, what are the experimental limitations? Where should the theory go next? Where does the future lie? Could clusters one day become the smallest devices or be used to make the smallest devices? Could one manipulate cluster isomers for the production new materials and nano-structures? What is the difference between a cluster and a bio-molecule or a virus? Could the molecules as complex as proteins or DNA and their functions be understood on the basis of the classical mechanics or one ultimately needs to invoke the quantum theory? What are the principles of the matter self-organization, self-assembling and functioning on the nanoscale? We merely mention such intriguing questions in this paper, but we hope that at least some of them will be resolved during the future development of *ab initio* and model approaches in studying structure and properties of ACs and complex molecules.

Acknowledgments

We dedicate this article to Professor Mohamed El Naschie on occasion of his 60th birthday with best wishes for a fruitful scientific future and personal happiness.

References

[1] J.P. Connerade, A.V. Solov'yov and W. Greiner, *Europhysicsnews* 33: 200, 2002.

[2] C. Guet, P. Hobza, F. Spiegelman and F. David (eds.), *NATO Advanced Study Institute, Session LXXIII, Summer School "Atomic Clusters and Nanoparticles"*, Les Houches, France, July 2-28, 2000, EDP Sciences and Springer Verlag, Berlin, Heidelberg, New York, Hong Kong, London, Milan, Paris, Tokyo, 2001.

[3] J.P. Connerade and A.V. Solov'yov (eds.), *Latest Advances in Atomic Clusters Collision: Fission, Fusion, Electron, Ion and Photon Impact*, World Scientific, London, 2003 (in print).

[4] A.V. Finkelshtein and O.B. Ptizin, *Physics of ptoteins*, University, Moscow, 2002.

[5] I.A. Solov'yov, A.V. Solov'yov, A. Koshelev, A. Shutovich and W. Greiner, *Phys.Rev.Lett.* 90: 053401, 2003.

[6] I.A. Solov'yov, A.V. Solov'yov and W. Greiner, *Phys. Rev.* A65: 053203, 2002.

[7] A.G.Lyalin, I.A. Solov'yov, A.V. Solov'yov and W. Greiner, *Phys. Rev.* A67: 0632XX, 2003.

[8] A.G. Lyalin, A.V. Solov'yov and W. Greiner, *Phys. Rev. A* 65: 043202, 2002.

[9] A. Matveentsev, A.G. Lyalin, I.A. Solov'yov, A.V. Solov'yov and W. Greiner, *Int.J.Mod.Phys.* E12: 81, 2003.

[10] I.A. Solov'yov, A.V. Solov'yov and W. Greiner, A. Koshelev and A. Shutovich, *International Meeting "From Atom to Nano structures"*, Old Dominion University, December 12-14 (2002), Norfolk, Virginia, USA (2002), American Institute of Physics, editors Jim Mc Guire and Colm T. Whelan, AIP Conference Proceedings, 2003, in print.

[11] W.A. de Heer, *Rev. Mod. Phys.* 65: 611, 1993.

[12] C. Bréchignac, J.P. Connerade, *J.Phys.B: At. Mol. Opt. Phys.* 27: 3795, 1994.

[13] H. Haberland (ed.), *Clusters of Atoms and Molecules, Theory, Experiment and Clusters of Atoms*, Springer Series in Chemical Physics 52, Berlin, Heidelberg, New York, Springer, 1994.

[14] U. Näher, S. Bjørnholm, S. Frauendorf, F. Garcias and C. Guet, *Physics Reports* 285: 245, 1997.

[15] W. Ekardt (ed.), *Metal Clusters*, Wiley, New York, 1999.

[16] S.Sugano and H.Koizumi, *Microcluster Physics*, Second Edition, Springer, Berlin, Heidelberg, London, 1998.

[17] H.W. Kroto et al., *Nature* 318: 163, 1985.

[18] J.M. Eisenberg and W. Greiner, *Nuclear Theory*, North Holland, Amsterdam, 1987.

[19] L.G. Gerchikov, P.V. Efimov, V.M. Mikoushkin and A.V. Solov'yov, *Phys.Rev.Lett.* 81: 2707, 1998.

[20] J.P. Connerade and A.V. Solov'yov, *Phys. Rev.* A66: 013207, 2002.

[21] I.A. Solov'yov, A.V. Solov'yov and W. Greiner, *XXIII International Conference on Photonic, Electronic and Atomic Collisions, Abstracts of Contributed Papers*, Stockholm, Sweden, 23-29 July 2003.

[22] Lord Rayleigh, *Philos. Mag* 14: 185, 1982.

Why are Probabilistic Laws Governing Quantum Mechanics and Neurobiology?

by Helmut Kröger

Abstract

We address the question: Why are dynamical laws governing in quantum mechanics and in neuroscience of probabilistic nature instead of being deterministic? We discuss some ideas showing that the probabilistic option offers advantages over the deterministic one.

1 Overview

Neuroscience is part of biology. In biology, Darwin's theory of evolution of species - which corresponds to the standard model in elementary particle physics - is based on two principles: The system undergoes small changes - encoded in the genes - which are random. Then there is competition (fighting, survival of the fittest) which means selection and which eventually may lead to the emergence of new species. As time evolves, biological evolution has generated forms of life starting from single cells, amoebia and eventually producing large animals (dinosaurs, whales). To use a modern term, complexity has increased. One may say that the random change is the motor which drives evolution, and selection takes care that complex forms of life emerge.

If we look at the evolution of the universe, the creation of elementary particles, the formation of nuclei (nucleosynthesis), the formation of atoms, and the formation of macromolecules like proteins, we notice that there is increasing complexity observable at different levels of length or energy (physicists often talk about scales of length or scales of energy, like Planck length, Λ_{QCD}, binding energy of proton, binding energy of ^4He, binding energy of atoms, of molecules etc.). Like in Darwin's theory, the concept of randomness is present in the evolution of the universe. For example, in the big-bang standard model of cosmology, the universe starts out from extremely high temperature where the laws of thermodynamics are believed to be valid. Thus the universe is described in terms of statistical field theory, where fluctuations occur in a random manner. During the early expansion phase, the inflationay model assumes that some tunneling transition has occured, which is of probabilistic nature.

The concept of probability and chance can be viewed as a motor driving the evolution of the living organisms but also the evolution of our universe. The essential difference between both is the mechanism of emergence of complexity.

The problem of emergence of complexity in the universe is related to the well known fine-tuning problem in nature: It means that there are some constants of nature the values of which necessarily have to lie in quite a small window, otherwise the complex universe as we see it today would not exist. Examples are windows for quark masses, windows for the cosmological constant, windows for the dipole moment of water and windows for a lot of other quantities. The question why such constants of nature take on values in such windows is an unresolved puzzle. Recently, this puzzle has become entangled with the observation that some constants of nature - like the fine structure constant α - have varied in time during the evolution of the universe [1]. The subject of this article is the question: Why does nature use the concepts of probability and chance in contrast to the alternative of determinism?

2 Introduction

We start out by asking the question: Are the basic laws governing in neurobiology of probabilistic or of deterministic nature? What if we ask the same question in quantum physics? The answer is well known for quantum mechanics: The laws are of probabilistic nature. In neurobiology one has not such a clear cut answer. But there is much evidence in favor of stochastic behavior. For example, there is irregular spiking behavior of cortical neurons in vivo. There is noise in synaptic transmission. Noise plays a role in the working of ion channels in the neural membrane. The reader might wonder why we are going to treat such different topics as quantum mechanics and neuroscience on the same footing, although it is generally believed that quantum mechanics is not needed to understand the working of the brain or of neuroscience in general. (An opposite view has been taken by Penrose [2]). The reasons are, first, that the typical length scales of Q.M. are similar to those playing a role in the working of a neuron. Secondly, the answers we are offering have something in common for Q.M. and neuroscience.

In the following we will focus on the question: Why are those laws of stochastic and not of deterministic nature? The reader might wonder in the first place: What makes us ask this question? Like in a detective story, asking the motives of a suspect often helps to trace the history of the crime and solve the murder puzzle. Likewise we hope that those questions will help to better understand neuroscience. Of course we can not give an answer to the main question. However, we will propose a tentative answer in the sense that it is favorable for nature to use the probabilistic option. We will present arguments in support of this thesis.

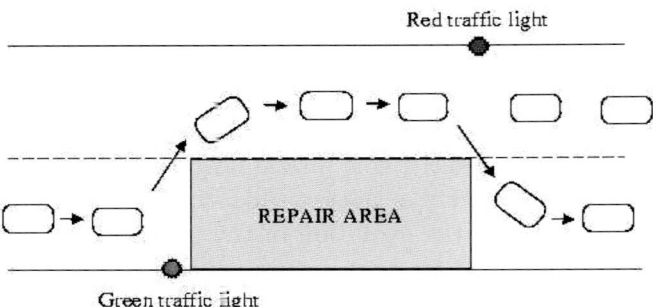

Fig. [1a] Highway traffic one lane in repair

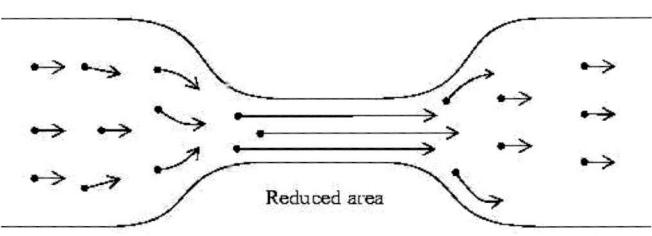

Fig. [1b] Crude oil in pipeline

Figure 1: Schema of flow of car traffic on highway (a) compared to flow of oil particles in pipeline (b).

3 Tentative Answer

The tentative answer which we propose for both Q.M. and neuroscience is: Nature has chosen the probabilistic option, because it offers the following advantages:

Quantum mechanics:
(1) Architectural simplicity and efficiency.
(2) Algorithmical simplicity.
(3) Cost efficiency.
(4) Repair efficiency and robustness.

(5) Infinite lifetime (in principle).

Neuroscience:
(1) Efficient (and may be the only) strategy for adaptive learning.
(2) Random connections are part of the small-world and scale-free network achitecture, which both were shown to be advantageous in nature.

Before entering into discussion, let us make a brief historic note. According to scientific evidence Q.M. and also neurobiology at the level of individual cells is ruled by stochastic laws. On the other hand, we know that the macroscopic world is described by classical physics, which is of deterministic nature. Historically, in the era of renaissance and rationalism, scientists believed that the whole universe could be described in terms of deterministic laws. With the advent of quantum mechanics, and the concept of probability involved, many people and in particular philosophers had great difficulties to accept that. Even Einstein believed in deterministic laws underlying quantum mechanics: "...Gott würfelt nicht..." (God does not role the dice).

This development of view and thought is quite natural, however. In the evolution of humans over the last 3 million years, man has investigated nature at the macroscopic scale, i.e. at the scale of resolution of the bare eye and length scales between resolution of touching sense to walking distance. Nature at the scale of atoms or neurons was not accessible to man until, say the last century. So mankind described nature in terms of classical physics. The interesting point is: When it became clear that probabilistic laws are at work at the length scale of Q.M., how did that change scientific thought? One direction was (Einstein): Shouldn't quantum mechanical laws be explained in terms of underlying deterministic laws? This direction corresponds to explain nature at the microscopic level by laws describing nature at the macroscopic level where nature is much more complex. So far there is no scientific evidence that this is possible.

Scientists also thought about the other direction: Can the deterministic laws of macroscopic physics be explained in terms of probabilistic laws of microscopic physics? This has been much more fruitful, as the success of statistical mechanics has demonstrated. Most scientists believe that the answer is yes! This direction corresponds to describe nature at the complex macroscopic level by reducing it to the laws of nature at the less complex microscopic level.

4 Example of Traffic: Deterministic Versus Stochastic Dynamics

Let us start by considering the following example well known from daily life, i.e., from the macroscopic world. Fig.[1a] shows automobile traffic on a two lane highway, with traffic passing an area of road work, with one lane being closed. At both ends of the repair area, traffic lights control the motion of traffic in both directions. At a given time traffic moves only in one direction. In this case, traffic is governed by a deterministic law (one can exactly predict, when the traffic light will change).

In Fig.[1b] we consider crude oil travelling in a pipeline, passing an area of reduced diameter. Oil molecules behave quite different from automobiles. They carry out Brownian motion with an underlying constant motion along the pipeline. This is a stochastic process. For an ideal (incompressible) liquid holds Bernoulli's equation $p + \frac{1}{2}\rho u^2 = $ const, where p is the pressure, ρ is the density and u the velocity of the liquid. Moreover, there is a simple equation of continuity

$$F_1 u_1 = F_2 u_2. \tag{1}$$

It means, when the liquid passes areas where the cross section of the tube is F_1 and F_2, respectively, then the velocity accomodates such that the product of velocity and cross section is constant. Consequently, the oil before and after the area of reduced diameter travels with the same speed. When the molecules reach the reduced section of the tube, particles will collide, but they begin to move faster. Comparing scenario (a) with (b), we find that in scenario (a) the traffic moves slower compared to when the repair area is absent. In scenario (b), the oil moves faster in the narrow zone and after having passed the narrow zone it moves with the same velocity as before that zone.

This is a first example showing that a system with underlying stochastic dynamics wins over a similar system with deterministic dynamics. This example demonstrates point (1): The traffic light supervision system is much more complicated than the pipeline, where a supervision mechanism is absent. Secondly, it is relatively slower. Of course scenario (b) holds only in a certain window of parameters. It depends on viscosity, pressure, velocity, Reynolds number etc. Under extreme conditions there is a transition to turbulence and the flow becomes jammed.

This example hints to some understanding why stochastic behavior wins over deterministic behavior. The oil molecules (although being large molecules) are objects of much simpler structure than cars. In particular,

Space Time Physics and Fractality

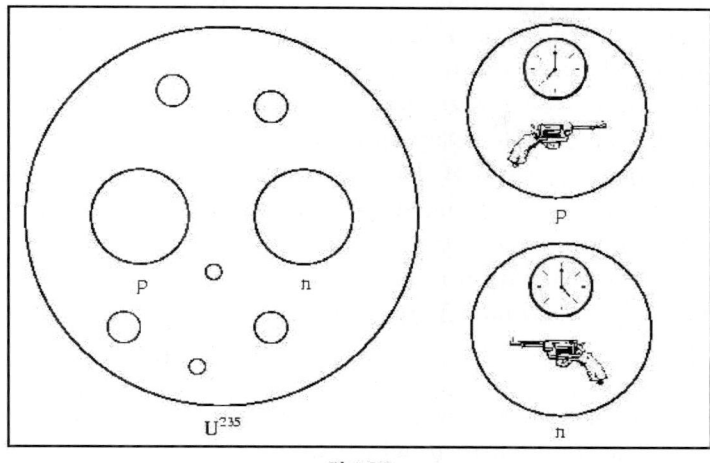

Fig. [2]

Figure 2: Scenario of deterministic decay of ^{235}U nucleus.

when undergoing collisions, the oil molecules come out undamaged, while cars (being complicated complex objects) would undergo heavy damage. So we formulate this observation: Microscopic objects of low-level complexity show stochastic behavior, while macroscopic objects of high-level complexity follow deterministic behavior. But aren't there counter examples to this claim? Let us consider the motion of ants. They apparently move very much like a random walker. And aren't ants beings of high complexity? Yes, they certainly are. However, their motion is not purely random. For instance when several ants want to enter into the ant hill at the same time, they do it in an orderly manner (more like the cars than the oil molecules) [3].

5 Probabilistic Nature of Quantum Mechanics

5.1 Single particle events in scattering

In Q.M. the occurrence of single events is purely random. For example, consider the emission of a photon from a light source. The time instant of emission can not be predicted. Why is it a random event? One can argue that, on the contrary, if it were not random or probabilistic, but deterministic, it would lead to contradictions in Q.M. As example let us consider Young's double slit experiment showing an interference effect of electrons. There is a source emitting electrons (like such existing in a television tube. The emission of electrons from a metal surface can be achieved by heating the metal). The electrons impinge on the screen A which has two holes. Some electrons get stuck. Some pass through the holes. Those which have passed continue and move on towards the second screen B. Some of those end up in the detector and are counted. The whole setup can be considered as a scattering experiment, where the beam consists of electrons and the target is represented by the screen A with two holes. In doing such experiment and counting the number of electrons (intensity) in the detector as a function of position x of the detector, one observes a curve with a maximum at the center $x = 0$. However there are several lowerside maxima. Also there are several minima, corresponding to intensity zero.

This is an interference pattern. Why is it called an interference pattern? Because it gives the same pattern, which would have been obtained in studying the behavior of water waves. Suppose one creates spherical water waves by periodically exerting pressure at the same place on the surface of water. These water waves propagate and arrive at the screen A (with holes positioned such that half of their opening is above water). Each hole creates a new spherical wave, which both propagate. Those interact with each other, creating a wiggly surface of minima and maxima, which can be observed in the detector.

So it turns out that electrons are behaving just like water waves. That is why quantum mechanics has historically been dubbed as wave mechanics. However, there is one essential difference: Water waves create this interference pattern only when secondary spherical waves are created at the holes in coincidence. That means the time instant when the spherical wave of the source arrives at hole number one is not very different from the instant when it arrives at hole number two. This is not necessarily so in quantum mechanics. First, when doing the quantum mechanical experiment with electrons, one can verify experimentally, that it is actually one electron at a time which passes a hole. And this electron passes either the upper hole or the lower hole at a time. Now comes the surprise. One can tune the electron source such that it emits an electron at a time with very long silent intervals in between. Nevertheless one observes that each electron contributes to build in the detector the interference pattern. With what did the electron interfer? If we would interpret the electron as an object obeying the laws of classical physics, there would be nothing the electron could interfer with, hence it hardly could generate an interference pattern. On the other hand, considering the electron obeying the laws of quantum mechanics, it is described by a wave function. The wave functioon can be decomposed in a part corresponding to the passage by hole

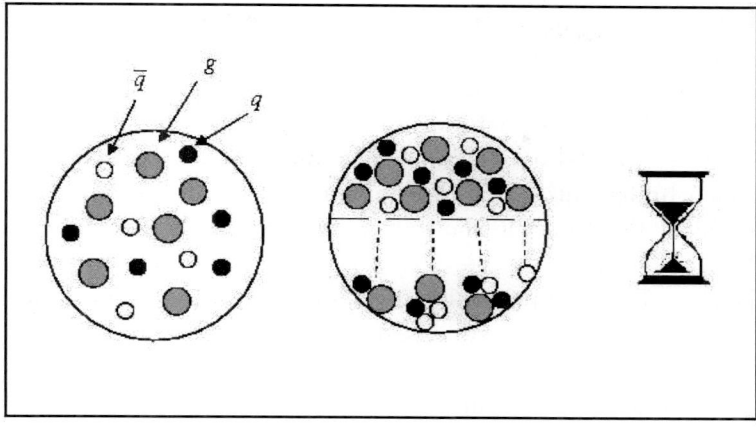

Figure 3: Schema of sand-clock of hadron, where quarks and gluons represent the "grains of sand".

one and a part corresponding to the passage of hole two. Then the interference pattern can beobtained by interference of the two pieces of wave function. The wave function $\psi(x)$ has an interpretation as probability amplitude. The probability itself is given by $P(x) = |\psi(x)|^2$ (see sect. 7.2). In conclusion, the scattering process of an electron from screen with two holes leads to a conflict, when adopting the classical physics, i.e. deterministic, point of view to explain the observed interference pattern.

5.2 Probabilistic decay of radio nuclei

Another similar example for the probabilistic nature of quantum mechanics is the decay of radio nuclei. E.g., let us consider an Uranium atom ^{235}U. It decays via certain pathways into a number of decay products, each decay channel being associated with a certain average life time. The question is: Why does it decay? What is the underlying mechanism? Of course we do have a quantum mechanical explanation for the decay. But we do not know of any deterministic law which tells us for a given particular ^{235}U atom, when exactly it will decay. There is general consensus that such law does not exist.

Why is there no such deterministic law? Let us put ourselves in position of the creator of an ^{235}U atom: Can we conceive a deterministic mechanism for the ^{235}U atom to have the decay properties observed in nature? How could that look like? The Uranium atom ^{235}U (see Fig.[2]) is built from protons and neutrons (the total number of which is 235). Let us consider the following scenario: Suppose in the interior of each proton and neutron there is a clock and a loaded gun (similar to Schrödinger's cat paradoxon). At a certain time, the moment of decay, the alarm rings. The alarm is connected with the gun. The gun fires a bullet onto some partner nucleon, giving it some momentum, which is sufficient to overcome the potential barrier and the atom decays.

Of course every child knows that this is fiction! But we ask: Why is this not realized in nature? The answer is: It would be way too complicated! The alarm clock and the gun are both objects of a much higher level of complexity then the proton and the neutron. So we ask: Can we do better and conceive a deterministic mechanism which is more simple (comparable to the level of complexity of the proton and the neutron)? Note that the most complicated object involved is the clock: Atomic clocks exist, which measures the oscillations of a Cs-atom. It requires a very complicated experimental setup in the laboratory. The alarm clock and the gun are objects of a much higher level of structure and complexity than the proton and the neutron. Note, that the ^{235}U atom or similar atoms may have quite a long life time (compared to intrinsic energy or time scales). If there would be an internal clock, it would need a memory device to store (count) the time units. It should be able to memorize time for quite a long time! Such a device would have a complex structure.

Can we construct a clock which is precise but no more complicated than an atom? In the Cs atomic clock, the Cs atom is not complicated, but the measuring apparatus is! Measurement means interaction. So we need an interaction. Let us consider the nucleon spin. The spin of a nucleon is aligned with the magnetic field if an external strong magnetic field is applied. The pointer of the clock corresponds to the direction of the precessing spin. But again measuring the direction of spin at a fixed time requires an apparatus much more complicated than a single atom!

We can imagine another possibility: It is well known that the nucleon has a substructure: Quarks and gluons. Imagine the "clock" to work like a sand-clock, but instead of sand grains utilizing quarks and gluons (see Fig.[3]). However, the motion of "sand grains" is more like a stochastic than a deterministic process. One could conceive many other possibilities of clocks and for the decay process. We claim that either the clock is much more complicated than the atom or the clock is based on a stochastic process.

Let us see, on the contrary, how the decay can be understood in a quantum mechanical, i.e. probabilistic, model. Consider Fig.[4]. There is a ground state and an excited state, represented by absolute and relative minima of the potential. The excited state is unstable. When the potential barrier is sufficiently high, the excited state is long living. In the limit of an infinite barrier height, the excited state becomes a stable state. The probabilistic laws of quantum mechanics allow for a transition from the excited state to the ground state (similar to tunneling). Note, that there is no additional mechanism necessary! The effect comes from the probabilistic nature of the wave function. Remark: A similar situation prevails, when considering a high lying excited state of, say a H atom (Rydberg states) with an excitation transition induced by photo absorption (laser) and de-excitation by photo-emission. The transition occurs due to the interaction with an electromagnetic field. Lesson: Q.M. predicts the transition probability and the decay rates. But it *does not* provide a deterministic statement of the precise decay time of a particular atom. The trade-off is: Q.M. probabilistic laws are much simpler then a deterministic law would have been!

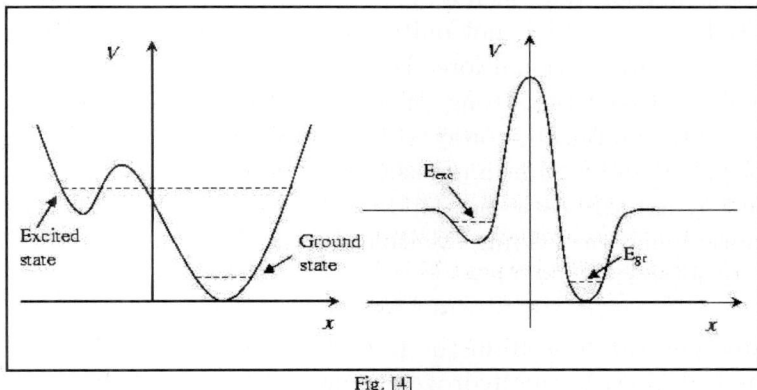

Fig. [4]

Figure 4: Potential model for Uranium decay as a transition from an unstable excited state to a stable ground state.

5.3 What if laws of physics at the atomic scale were deterministic?

In the real world, we believe that nature at atomic scales is described by quantum mechanics. There is a wave function $\psi(\vec{x}, t)$ which is a complex number (for any given value of $\vec{x}$ and t). It has a probabilistic interpretation.

$$P = |\psi(\vec{x}, t)|^2 \Delta V \qquad (2)$$

gives the probability to find a particle, obeying the Schrödinger equation with a wave function ψ, in a space volume ΔV around the position $\vec{x}$ at time t.

Since the advent of quantum mechanics, we can explain the discrete levels of bound states in atoms, which classical physics, based on deterministic laws, failed to explain. This was the first great success of quantum theory! We ask: What if an atom would be a classical object, governed by deterministic laws? Firstly, one can look from the point of view of classical chaos. It has been known since Poincaré that a classical 3-body system may be a chaotic system. Thus our solar system may not be stable. Numerical solutions of orbits by Sussman and Wisdom [4], Laskar [5] and recent analytic calculations by Murray and Holman [6] show that the jovian planets Jupiter, Saturn, Uranus and Neptune are chaotic, with an estimated Lyapunov life time of 10^7 years. If we consider a 3-body system at the atomic scale, an example would be the deuterium atom (heavy water) consisting of a proton, a neutron and an electron. Under the assumption that such a system would be ruled by classical laws, it would likely be chaotic (it is not quite the same as the celestical 3-body system, because the gravitational force is $F \propto 1/r^2$, which is if the same type as the Coulomb force, but the strong force between proton and neutron is not of this type). More complicated objects like large molecules with a complicated binding mechanism are quantum mechanically stable. If such a system would be governed by deterministic laws, it would quite likely lead to collisions and strong chaotic behavior, leading eventually to decay, which means a short life time of such objects.

Secondly, one can look from the point of view of classical electrodynamics. In classical physics, the hydrogen atom is a 2-body system composed of a heavy proton and a light electron, both carrying electric charge, and the electron orbits around the proton. The orbiting electron follows a curved trajectory, which means the particle undergoes acceleration. Classical electrodynamics predicts that this causes radiation, which goes with a loss of kinetic energy. As a consequence, the atom would not be stable and collapse rapidly. This effect is much stronger than the chaotic dynamics effect. The electrodynamical radiation loss and its prediction of unstable atoms was an import incentive for the invention of quantum mechanics.

As a result, if an atom would obey the laws of classical physics, it would be an unstable object. No macromolecules (proteins) would exist in nature. DNA would not exist, and hence no organic life.

5.4 Existence of stable atoms and probability

We want to show that the existence of stable atoms can be traced back to the concept of probability in quantum mechanics. A stable atom means that there are discrete energy levels $E_0 < E_1 < E_2...$, with gaps $\Delta E_i = E_i - E_{i-1} > 0$. Once the atom occupies one of those states, it can stay in this state forever (if there are no interactions with any other atoms or any electromagnetic field). In classical mechanics as well as in Q.M., the Hamiltonian is given by kinetic energy plus potential energy,

$$H = T + V. \tag{3}$$

We claim that a discrete spectrum is possible only when T and V do not commute. Mathematically, this means

$$TV - VT \equiv [T, V] \neq 0. \tag{4}$$

Physically, this means that kinetic energy and potential energy can not be measured simultaneously.

Theorem. $H = T + V$ has a discrete spectrum only if $[T, V] \neq 0$.
Proof (heuristic). Let us assume that T and V commute,

$$[T, V] = 0. \tag{5}$$

There is a mathematical theorem, stating that if two operators commute, then they can both be diagonalized in a common basis. The kinetic energy operator $T = \frac{\vec{p}^2}{2m}$ is diagonalized in a momentum basis

$$T|\vec{p}> = \frac{\vec{p}^2}{2m}|\vec{p}>. \tag{6}$$

Hence also V must be diagonal in this basis,

$$V|\vec{p}> = v(\vec{p})|\vec{p}>. \tag{7}$$

The Schrödinger equation implies

$$\left(\frac{\vec{p}^2}{2m} + v(\vec{p})\right)|\vec{p}> = E_p|\vec{p}>. \tag{8}$$

Thus we find for the spectrum

$$E_p = \frac{\vec{p}^2}{2m} + v(\vec{p}). \tag{9}$$

Making the reasonable assumption that the potential is at least a piece-wise continuous function, one obtains that the spectrum E_p may have a multi-band structure, but certainly is not compatible with the structure of individual discrete levels with finite gaps, which proves the theorem.

In a one-body system one has

$$T = \frac{\vec{p}^2}{2m}, \quad V = V(\vec{x}). \tag{10}$$

Eq.(4) is equivalent to

$$[\vec{X}, \vec{P}] \neq 0. \tag{11}$$

Actually,

$$[X, P_x] = i\hbar, \quad \text{same for y- and z-component} \tag{12}$$

is the fundamental commutator relation between position and momentum in Q.M. The latter equation is closely related to Heisenberg's uncertainty relation

$$\Delta X \, \Delta P_x \geq \frac{\hbar}{2}, \quad \text{same for y- and z-component} \tag{13}$$

being fundamental property which has been experimentally observed. Actually, Eq.(13) can be derived from Eq.(12) [7]. Here ΔX is defined by

$$\begin{aligned} \Delta X &= \sqrt{<\psi|(X-\bar{X})^2|\psi>} \\ &= \int d^3x (x-\bar{X})^2 |\psi(\vec{x})|^2 \\ &= \int d^3x (x-\bar{X})^2 P(\vec{x}), \end{aligned} \tag{14}$$

where $\bar{X} = <\psi|X|\psi>$ denotes the mean value of X in the state ψ. ΔX is the root mean square deviation or variance of X for a probability density distribution given by $P(x) = |<\vec{x}|\psi>|^2$. ΔP is defined correspondingly. As can be seen, both ΔX and ΔP depend on the particular wave function ψ. Heisenberg's uncertainty relation can be interpreted as an upper bound for any wave function ψ on the product of variances of position and momentum. This shows how the concept of non-commuting operators leads to the notion of mean and variance, both concept from probability theory. This ends our discussion on the relation between stable states and probability in Q.M.

6 Implication of Probability on Geometry

Above we have discussed the relation between probability, Heisenberg's uncertainty relation, and the commutator between the position and momentum operator. The occurrence of non-vanishing variance means that there are quantum fluctuations. A well known example are zero-point fluctuations for the ground state energy, which implies that classical and quantum ground state energy differ. Another example is the propagation of a quantum particle. Feynman's path integral describes the propagator as a sum of weights $exp[iS(x(t))/\hbar]$, where S is the action and $x(t)$ is a path from the starting point to the end point of propagation. Infinitely many such paths contribute to the propagator. These paths can be viewed as fluctuations around the classical trajectory. The propagator can be viewed as the wave function corresponding to the specific initial condition that the particle is located initially at the starting point of propagation. This connects quantum fluctuations to probability of the wave function.

Now when we speak about geometry in quantum physics, we do not mean the standard coordinate system of position $\vec{x}$ and time t. But we mean to talk about geometry closely related to the dynamics of the system. This idea is basically Einsteins old idea to express forces of gravitation in terms of the geometry of space time. Can we do something similar in quantum physics? In an attempt to do this, one can consider paths occuring in the path integral of the propagator. We introduce a geometry in the space of paths by introducing a distance of paths. We can do this in the following way: Say we have two paths, $x_1(t)$ and $x_2(t)$. There are two corresponding weight factors $exp[iS(x_1(t))/\hbar]$ and $exp[iS(x_2(t))/\hbar]$. We say that two paths are equivalent, if their corresponding weight factors are identical, i.e. their corresponding action is identical. We can define the distance bewteen two paths by

$$d(x_1(t), x_2(t)) = |S(x_1(t)) - S(x_2(t))| . \tag{15}$$

It is intuitively plausible that paths with small action S give the dominant contributions to the propagator (otherwise for a path with large action, a little change in the path would lead to strong oscillations and eventually cancel out all those terms). This means also that paths with large fluctuations between neighbor time slices are unfavorable, because they contribute to a large kinetic term in the action. In the case where the potential is of confining type, i.e. $V(x) \to \infty$ when $|x| \to \infty$, then also large fluctuations, where the path deviates much from the classical path are unfavorable, because they give a large potential term in the action.

The above definition of geometry is purely classical so far: S is the classical action and $x(t)$ is an arbitrary (random) path. Consequently the metric is purely classical. This is changed if we select representative paths drawn from a distribution involving the weight factor $exp[iS(x(t))/\hbar]$. This is exactly what is done when computing the quantum propagator (in imaginary time) via Metropolis Monte Carlo. Then one finds the action of paths to follow a Gaussian distribution centered around a small value of action. Then the above metric gives some information about the distance of quantum mechanical important paths.

A different notion of geometry arises, when we look at the fluctuations due to the kinetic term of the action. A quantitative way to do this is by looking at the Hausdorff dimension d_H of an average path $x(t)$. This can be done by computing the expectation value $<L_{path}>$ of the length of paths averaged over all paths and weighted by $exp[iS(x(t))/\hbar]$. In a numerical simulation by Monte Carlo one measures $<L>$ versus the variance Δx. In the limit when $\Delta x \to 0$, one obtains a power law of the form

$$<L> \propto \Delta x^\alpha , \qquad (16)$$

The exponent α determines the Hausdorff dimension ($\alpha = 1 - d_H$). More details of numerical experiments can be found in Refs. [8, 9]. Numerical results show $d_H = 2$, for all local potentials confirming the analytic result by Abbot and Wise [10]. This means the Hausdorff dimension is not sensitive to the interaction. This is a somehow deceptive result, in the sense, that fractal geometry is not a useful tool for the purpose of a geometrical interpretation of quantum physics.

However, the above result is valid in flat Riemannian geometry, i.e. in absence of any gravitational field. The situation is different, when one considers quantum mechanics in curved space time, i.e. when a quantum particle propagates in the neighborhood of a massive stellar object like a neutron star or a black hole. One expects that in such situation the fractal dimension should differ from $d_H = 2$.

7 Quantum Mechanics Versus Neuroscience

7.1 Typical scales: Q.M. versus neuroscience

First of all we want to state clearly that neuroscience is generally considered as a field where the laws of classical physics apply and the laws of quantum

physics are not relevant. In order to get a picture let us look at some typical scales of energy, length and time, being characteristic in quantum physics versus neuroscience. In quantum physics the following quantities play a role in setting scales in atomic physics:
(1) Action: $\hbar = 6.58 \ 10^{-22} \ MeV \ s$. This is a fundamental constant of nature.
(2) Energy: E_0 the ground state energy of an atom. For the hydrogen atom, $E_0 = -13.6 \ eV$.
(3) Time: $\hbar/E_0$. This sets the time scale in dynamical processes, like a tunneling transition. For the hydrogen ground state $\hbar/E_0 = 0.48 \ 10^{-16} \ s$.
(4) Length: λ the de Broglie wave length of a particle. For a thermal electron of kinetic energy $1 \ eV$, one has $\lambda = 1.95 \ 10^{-8} \ cm$. Another typical length scale is radius of an atom, e.g., the Bohr radius of the hydrogen atom, which is in the order of 1 Angstrom.

In neuroscience the following scales play a role:
(1) Size of central nervous system 1 m.
(2) Size of a neuron 100 μm.
(3) Size of a synapse 1 μm.
(4) Typical length of dendrite 1 mm.
(5) Time: duration of action potential 1 ms.
(6) Firing rate 50 $Hertz$.
(7) Refractory period 1 ms.
(8) Oscillations: $\alpha, \beta, \gamma, \theta$ 5-80 Hertz.
Obviously typical length and time scales differ from atomic physics to neuroscience.

7.2 Different concepts of probability: Q.M. versus neuroscience

The concepts of probability in quantum mechanics and in neuroscience are different. In quantum mechanics probability is introduced as the interpretation of the absolute square of a transition amplitude or wave function. In neuroscience it shows up e.g. in the erratic (noisy) behavior of ions passing membranes via ion channels or in the diffusion of neurotransmitters in synaptic transmission. While the wave function obeys the Schrödinger equation, the diffusion of neurotransmitters obeys the diffusion equation. In order to make the distinction clear, let us consider the case of free motion (absence of any driving force or potential). Then the Schrödinger equation reads

$$-\frac{\hbar}{i}\frac{\partial}{\partial t}\psi(x,t) = -\frac{\hbar^2}{2m}\frac{\partial^2}{\partial x^2}\psi(x,t). \qquad (17)$$

On the other hand, the diffusion equation can be written as

$$\frac{\partial}{\partial t} P(x,t) = D \frac{\partial^2}{\partial x^2} P(x,t). \tag{18}$$

Here $P(x,t)$ denotes the probability to find the particle at position x and time t. D detotes a constant, which characterises the diffusion process. Mathematically one observes a great similarity between both equations. Actually, the Schrödinger equation goes over to the diffusion equation under the transformation

$$\begin{aligned} t &\longrightarrow -it \\ \frac{\hbar}{2m} &\longrightarrow D. \end{aligned} \tag{19}$$

While the second equation is a mere scale transformation, the important difference is due to the complex number i. As a consequence the laws of adding probabilities in quantum mechanics and in a diffusion process are totally different. This point has been discussed in a most clear and lucid way by Feynman and Hibbs [11]. The mathematical rule of adding probabilities in Q.M. is the following: The basic entity is the wave function or probability amplitude ψ. The wave function at some position $\psi(\vec{x})$ is a complex number. The probability to find a particle in volume ΔV around position $\vec{x}$ is given by

$$\Delta P = |\psi(\vec{x})|^2 \Delta V . \tag{20}$$

In shorthand, the probability is related to the probability amplitude by

$$P = |\psi|^2 . \tag{21}$$

The prescription in quantum mechanics is: One does not add probabilities, but the probability amplitudes.

$$\begin{aligned} &\psi_1, \psi_2 \to \psi = \psi_1 + \psi_2 \\ &P_1 = |\psi_1|^2, \quad P_2 = |\psi_2|^2 \Longrightarrow \\ &P = |\psi|^2 = P_1 + P_2 + 2Re[\psi_1^* \psi_2] \neq P_1 + P_2 . \end{aligned} \tag{22}$$

In diffusion dynamics the rule is

$$P_1, P_2 \to P = P_1 + P_2 . \tag{23}$$

The physical consequence of the diffent laws of adding probabilities can be seen, for instance, in an interference experiment. In quantum mechanics, when a source emits electrons, which pass through a screen with two slits, a detector counts intensity (probability) which has a typical shape of maxima alternating with minima. The minima are due to destructive interference which is possible due to the term $2Re[\psi_1^*\psi_2]$. Contrary to that a particle obeying diffusion dynamics yields an interference pattern with a single maximum only (see Ref.[11]).

One may take a closer look at the law of adding probabilities in Q.M. and ask: Why is it that the wave function has to be complex? After all, some destructive interference might occur already, if ψ_1 and ψ_2 would be real but of opposite sign. Some indirect evidence is given by the Aharonov-Bohm experiment, where a magnetic solenoid is placed in Young's double slit experiment. Then the wave function reveals topological behavior (i.e. its phase distinguishes if the solenoid is interior or exterior to a closed loop formed by the classical trajectories of two particles going from the source to the detector but traversing different slits). Such behavior is possible only when the wave function is complex. Experimentally, such complex phase factor results in a lateral displacement of the interference pattern.

There is another plausibility argument showing the necessity of the complex unit i to occur in the wave function. Let us consider the time evolution in classical mechanics in a Hamiltonian system with a Hamilton H. It can be expressed by an operator $\exp[\tilde{H}t]$ by

$$q(t) = \exp[\tilde{H}t] \; q(t=0) \;, \qquad (24)$$

where $\tilde{H}$ denotes a Lie operator, the mapping of which applied to a phase space variable q is given by taking the Poisson bracket of q with the Hamiltonian H. The previous equation is similar to the evolution of the wave function under a (time-independent) Hamiltonian in Q.M.,

$$\psi(t) = \exp[-iHt/\hbar] \; \psi(t=0) \;. \qquad (25)$$

The difference between both equations is, first of all that $\tilde{H}$ and H operate in different function spaces. But more importantly, the wave function guarantees the conservation of probability, which is not the case in classical mechanics (where the Liouville measure is conserved). This means that $\exp[-iHt/\hbar]$ must be a unitary (or anti unitary operator). But H representing the observable energy is Hermitian. The factor i occurs a necessity when relating a Hermitian operator with a unitary operator of exponential form. The same argument applies when considering rotations, being also unitary operators, where the generators are the operators of angular momentum (or spin), which again are Hermitian.

8 Random Behavior in Neuroscience

Neurons are noisy [12]. First, there is noise in the ion channels. Second, a neuron in the visual cortex, when repeatedly stimulated, never responds in the same way, neither in time nor in amplitude [13]. Another example is synaptic transmission. When a spike arrives from the axon at the presynapse, it does not trigger with certainty a signal on the postsynaptic side. The liberation, propagation and the process of docking of neurotransmitter molecules is a random process. The neurotransmitter molecules are large molecules (proteins).

One may ask, if a single neurons acts noisy, how does the brain "know" about the presence of a stimulus? In the brain usually very many neurons respond to a stimulus, hence the brain can filter out a signal from the noise. This has a mathematical foundation in probability theory and statistics. The central limit theorem says that when taking the arithmetic mean of a large number N of random variables, the statistical error σ of this mean goes to zero like $1/\sqrt{N}$, i.e. behaves more and more deterministically. As a consequence, a phenomenological model describing the activity of the membrane potential of the neural cell, which involves thousands of ion channels, can be well described by a deterministic model, the Hodgekin-Huxley equations.

From this one gets the impression that noise seems to be a nuisance, like in many areas of science. However, there are indications that the brain may also take advantage of the presence of noise. An example is the mechanism of stochastic resonance, which serves as an amplification mechanism of weak signals. This and other examples of the role of noise will be discussed in the following.

8.1 When noise in neurons plays a constructive role

In physics, there are many examples, where noise plays a destructive role: Example: Line broadening of a spectral line, due to thermal oscillation of the atom which emits the light. Another example: Diffuse background light being emitted from populated areas, which disturbs astronomical observation in the night. On the other hand, there are phenomena in nature, which are due to a stochastic process or which are coupled to a stochastic process, and where the presence of noise plays a constructive role: As a result the signal-to-noise ratio is enhanced, or more generally there is creation of order out of disorder. For example, noise plays an important role in the mechanism of hearing in the ear [14]. We will discuss the following three mechanisms:
(1) Auto-criticality and the sand pile model,
(2) Self-organisation off thermal equilibrium: creation of order out of disorder.

Examples are the Belousov-Zhabotinski reaction, and Bernard convection. (3) Stochastic resonance and ergodicity breaking in systems with many valley structure.

8.2 Sand pile model

The sand pile model is a simple model for a system which keeps some parameter at its critical value [15]. It does so without external interference. It is a mechanism for self-organized criticality. It explains the $1/f$ noise observed in transport systems like resistors, the hour glass and luminosity of stars. The $1/f$ behavior reflects a critical state of minimally stable clusters of all length scales. The model has been related to the behavior earthquakes, forest fires, ecology, stock markets, and weather.

Imagine a pile of dry sand. Possibly remote memories from childhood on the construction of sand castles tell us that those construction used to decay under the influence of sun and wind into a lump of sand with a shape similar to a cone. Suppose we have such a cone. Then adding on top a few more grains of sand may trigger an avalanche of sand, such that as outcome a cone shape is restored. The point is that there is a critical value of steepness, which the sand pile tries to maintain. Now the avalanche is a stochastic process, which is needed to keep the system at its critical value of steepness.

Does such mechanism play a role in biology? First, an ant hill resembles very much a sandpile, it is conceivable that ants (European red would ant) in constructing the ant hill have do deal with critical stability. One may ask: Do ants "know" the critical steepness? Second, does such kind of mechanism of self-organized criticality play a role in the working of the brain? This is not known! However, one might speculate about this mechanism because parts of the brain are known to function quite autonomously (breathing, heartbeat).

8.3 Self-organisation off thermal equilibrium

Prygogine and collaborators were the first to propose and develop the idea that in nature processes off thermodynamical equilibrium occur where locally the entropy decreases leading to states of higher order. In biology this may lead to forms of life of higher order and complexity. A fine discussion of self-organisation off thermodynamical equilibrium, is given by Nicholis [16]. Well known examples are the Belousov-Zhabotinski reaction and Bernard convection. Let us consider the Bernard convection. Anyone who has been heating oil in a frying pan has had the chance to observe this phenomenon: At the beginning when the heat has just been turned on, the surface of oil is quiet and flat. After some time when the heat is sufficienly strong, one observes that the fluid of oil creates honey-comb like structures. They are quite stable. The physical reason is the difference of temperature between the pan and the upper surface of the liquid. It creates a circular motion, which manifests itself in an ordered structure. Due to the temperature difference this is an off-equilibrium process. It is evidently a random process. Its outcome has a non-random deterministic geometrical shape. Considering oil, pan and source of heat as a single system, this system manages to create order out of disorder (a signal out of noise) without external interference.

The Bernard convection occurs also elsewhere in nature, for example in the motion of air stream in the upper atmosphere. One may even speculate if the formation of galaxies, which is known to have voids and some regular structures [17] may have to do with the Bernard convection mechanism. The question is: Is such mechanism pertinent in the dynamics of the brain? This is also unknown! However, the brain is known to create spatially and temporally domains of order, where peviously there has been disorder. An example is the desease of epilepsy during an attack, where quite regular (orderly) firing patterns are observed in neural activity.

8.4 Stochastic resonance

The mechanism of stochastic resonance has been proposed to explain the recurrence of ice ages [18]. Reviews of the mechanism of stochastic resonance can be found in refs. [19, 20, 21]. Longtin [22] proposed that this mechanism plays a role in excitable systems like neuron models. Douglass et al. [23] demonstrated this mechanism at work in mechanoreceptor cells in crayfish. Collins et al. [24] showed that it enhances tactile sensation in man. Also it is present in the sensor neurons of the rat [25], as well as in crickets [26]. A recent review on more examples of stochastic resonance is given in ref. [27].

What is stochastic resonance? As a simple example consider in classical physics a particle moving in a double well potential. If the kinetic energy is insufficient to overcome the potential barrier, the particle is confined to stay in the same well all of the time. Now suppose one adds friction, also a periodic force which lowers one potential bottom and raises the other one and after one period goes in the opposite direction. Finally, one adds a random (noisy) force which helps the particle to overcome the barrier. Then an interplay between periodically changing potential, friction and random force creates a "resonant" motion of the particle from one potential bottom to the other.

How can such mechanism play a constructive role in neuroscience? Let us consider a neuron and the creation of an action potential. In simple terms the neuron can be viewed as a 2-state system: one state when it fires a spike, and the other state when it is quiet. The mechanical analogue is a system where the particle is in one or the other well. Adding a noisy force may help the mechanical system to more easily go over from one state to the other. Thus the presence of noise in the neuron may change its response to create an action potential. Such noise is present in excitatory synapses.

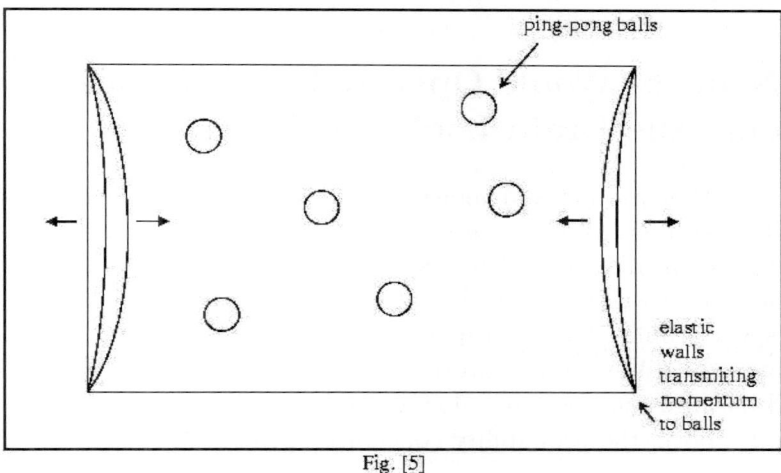

Figure 5: Schema of Brownian motion in macroscopic physics: Ping-pong balls in a closed box.

Another example where noise may play a constructive role, is the associative memory in the brain. The associative memory has been described by the Hopfield model. There is an energy surface of configurations (corresponding to an ensemble of neurons, each one being in the state of firing or quiescence). The Hopfield model is mathematically equivalent to a spinglass in condensed matter physics. Spin glasses exhibit a many-valley structure in the free energy (broken ergodicity). In the associative memory model, the bottom of a valley corresponds to a stored pattern, i.e. a piece of stored memory. Suppose one wants to go from one pattern of memory to another i.e. from one valley to another. Those who hiked in mountains know that this can cost quite a bit of energy. Such energy may be provided in form of some noise. Also when solving such spin glass models numerically, one adds noise to facilitate the migration through the whole phase space to search for the global inimum (algorithm of simulated annealing). May be the brain uses a similar method to switch from one memorized pattern to another pattern.

I summary, noise helps to overcome ergodicity breaking (not getting caught in a valley). This is likely to be important in the brain. Going from one valley to another in the brain means to go from one memorized pattern to another memorized pattern. In this picture the role of noise can be seen as a motor which helps to populate higher lying energy levels of a spin glass system, which corresponds in the associative memory to memorize or access memorized patterns located at higher values of the energy (cost) function.

9 If Neurons Would Operate Deterministically How Would the Brain Look Like?

Let us recall that atomic and molecular physics is based on the concept of probability (see sect. 7.2). Neuron dynamics shows stochastic behavior in the synaptic transmission and in the opening/closing of ion channels in the membrane. Both processes involve atoms and molecules. Thus the probabilistic aspect is inherited from quantum mechanics. On the other hand, in neuroscience the diffusion of neurotransmitter molecules has been successfully described by diffusion equations, being the limit of Brownian motion. As we pointed out above, the probability concept of quantum mechanics is different from that of Brownian motion.

We should note that neurotransmitter molecules are quite large molecules (proteins). We ask: Is Brownian motion a property of microscopic particles, obeying or can macroscopic particles also carry out Brownian motion? We suppose the answer is yes! As an example consider ping-pong balls in a large container with elastic walls transmitting momentum to the balls (see Fig.[5]). The mean free path should be much larger than the size of the balls. This is a question of density. Brownian motion requires a large number of particles to enter in collisions. In the synapses, the collisions occur between the neurotransmitter molecules and the liquid in the synaptic cleft.

So we come back to the question: If a neuron would not operate stochastically - described by Brownian motion - how could it operate? Could it work at all? First of all this means that a synapse would be physiologically different. One can speculate about chemically different neurotransmitter molecules, which would be of different size, leading to different collision rates and hence different diffusion properties. Or it would be conceivable that the noise would be frequency-dependent: Fast diffusion might correspond to a noise with small sigma (deterministic limit). Or one might consider the fluid in the synaptic cleft to be chemically different, e.g. like a colloid, where neurotransmitter molecules might get stuck. All this would lead to a different synaptic transmission behavior. Ultimately, is it possible that a biological synapse could work deterministically? I think the answer is no! The argument is of the same kind as for the quantum mechanical decay of a radio-nucleus: It would be too complicated!

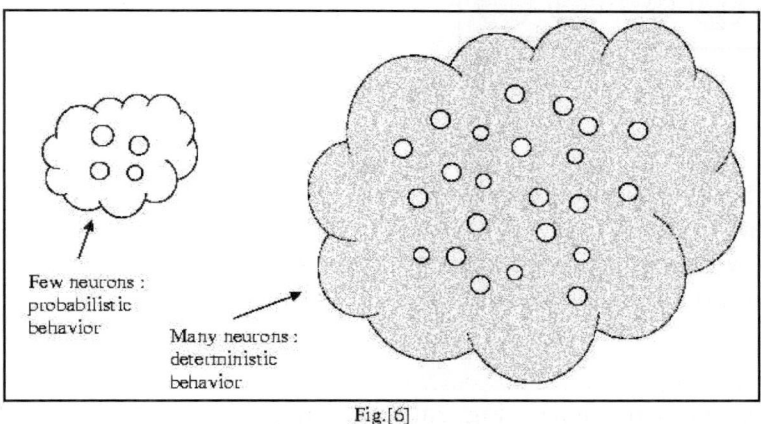

Figure 6: Activation of few neurons gives probabilistic response, while coherent activation of many neurons gives deterministic response.

9.1 Dynamic regimes where the brain operates deterministically

Let us consider as example the motoric system of the brain. When a fast response to a stimulation is required (an animal flees after an attack from a predator), then many cells fire together. A big muscle undergoes contraction. This requires a coherent effect of many neurons. The ensemble of motor neurons responds in a deterministic way. This situation is similar to the transition in physics from a few-body system to many body system. For a many-body system (at thermodynamical equilibrium) one can use statistical mechanics. Macroscopic observables behave classically (Central Limit Theorem). This is realized also in the brain (see Fig.[6]). How can we tell in which mode the brain is working - probabilistic or deterministic? A schematic plot of possible

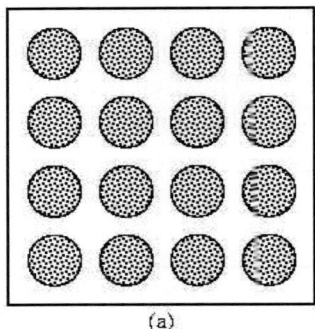 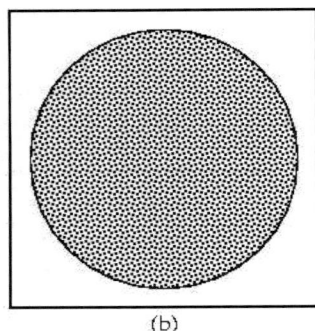

Figure 7: Schema of organisation of brain. (a) Many small areas with a few neurons interacting weakly: brain works in a probabilistic mode. (b) One large area with many neurons interact strongly (coherent activation): brain operates in a deterministic mode.

scenarios is shown in Fig.[7]. Fig.[7a] shows are many small areas with a few neurons. If the small areas interact only weakly then this part of the brain likely works in a probabilistic mode. In Fig.[7b] there is one large area where the neurons interact strongly then this part of the brain operates coherently (at least for some time). Then the brain is in a deterministic mode. The question is then: What exactly is a strong or weak interaction? Such questions have been addressed in neural network models. Noest [28] has shown that neural networks can form domains from restricted range interactions. Recently an interesting answer has been given by the proposal of small world networks and scale-free networks (see sect. 11). It has been shown that the small world architecture of neural nets - characterized by strong local clustering and a few short links to distant neural nodes - is capable to yield a fast coherent response [29]. The issue of the architecture of the brain raises further questions about long range order, phase transitions, order parameters etc. in the brain. Not much is known about this.

10 Learning

Here we want to discuss the process of learning, retrieval of learned information and their neurophysiological basis using chance and probability.

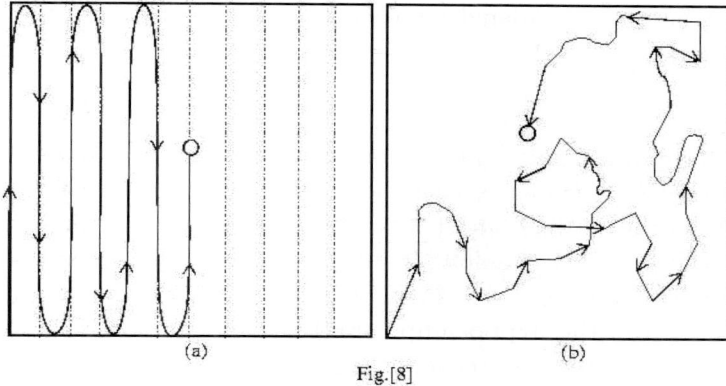

Figure 8: Possible paths in a search: (a) regular path versus (b) random paths

10.1 Cellular basis of learning and formation of memory

Learning and memorizing are mental abilities found in humans and mammals, but also in animals of simpler organisation having only a number of neurons in the order of 10 to 10^5. An example is the sea-snail (Aplysia californica). This animal shows a simple forms of learning: habituation and sensitisation[30]. Habituation is observed, when the animal is exposed to some stimulation of its breathing organs. It responds by a retraction-reflex of its gills. When repeating this simulation 10 to 15 times, the retraction-reflex does no longer show up. Even after one hour this reflex occurs only in a much weaker form. The animal has undergone some habituation. Sensitisation is the opposite effect, where some response of the animal becomes stronger. The molecular mechanism of short term habituation and also sensitisation has been localized to occur in the synaptic transmission. It is based on a change in the current of Ca^{++} ions passing through ion channels in the membrane of the synaptic boutons of sensory neurons. The concentration of Ca ions in the boutons controls how many synaptic vesicles will deliberate their neurotransmitter molecules into the synaptic cleft (space between pre-synaptic and post-synaptic neuron), after an electric action potential has arrived at the presynaptic bouton. In the process of habituation, the number of free Ca ions decreases in the boutons, resulting in reduced flow of neuro-transmitter molecules. Eventually the reflex vanishes completely. On the contrary, in the process of sensitisation, the flow of free Ca ions into the synaptic bouton is increased, resulting in an amplified flow of neurotransmitter molecules. The molecular basis is more complicated, requiring the interaction with an interneuron (and a chain reaction involving Serotonin, Adenylat-Cyclase, cyclic Adenosinmonophosphate, Protein-Kinase).

A higher form of learning and memorisation is found in mammals and humans. It is based on a change (plasticity) in the synaptic transmission. One makes a distinction between short term potentiation (STP) and long term potentiation (LTP). LTP is a long-lasting increase in the amplitude of the synaptic response following brief, high-frequency activity of a synapse. LTP was first described in the hyppocampus by Bliss and Lomo [31]. Let us consider as example of LTP the Schaeffer collateral CA1 pyramidel cell in the hyppocampus [32]. A change occurs in the synapse on the post-synaptic side. The origin is an interaction between ionotropic glutamate receptors, AMPA (2-amino-3-propanonic acid) and NMDA (N-methyl-D-aspartate). First the AMPA and NDMA ion channels are closed. When the post-synaptic membrane becomes depolarized, the neurotransmitter interacts with AMPA and NMDA ion channels. Then the AMPA channels open to create an excitatory post-synaptic potential. But the NMDA channel is blocked by Mg^{2+} ions. Only when the post-synaptic membrane is very strongly depolarized, the Mg^{2+} ions are liberated and then NMDA channels open. The NMDA channel allows Na^+, K^+ and Ca^{2+} ions to pass. The Ca^{2+} ions induce a number of biochemical reactions increasing the efficacy of the synapse. Also there are hints that this LTP process is accompagnied by a release of nitric oxide (NO), which plays the role of a signal being sent back to the pre-synaptic terminal inducing additional neurotransmitter release [33]. The NO molecule was found to enhance transmitter release only if it arrives in coincidence with activity in the pre-synaptic neuron.

Now we ask: Where in those processes of learning does chance and probability play a role? In the case of habituation and sensitisation, the process involves the diffusion of Ca ions through the membrane ion channels. This is a stochastic process. Secondly, controlled by the increase or decrease of Ca ions, there is an increased or decreased number of neurotransmitter molecules, diffusing through the synaptic cleft. Again this is a stochastic process. Similarly, in the case of LTP, there are two occasions, where randomness plays a role. First, it is the diffusion of neurotransmitter and the docking at $AMPA$ and $NMDA$ receptors, which is a stochastic process. Secondly, also the retrograde signal of the NO molecule proceeds via some diffusion process, again subject to the laws of chance.

10.2 How and where is information stored in the memory?

We know that learning in humans and most likely also in animals can be distinguished as explicit and implicit learning. Explicit learning means to remember and recognize people or places. It involves the temporal lobe, the hippocampus (for short term memory) and the cortex (for long term memory). Implicit learning means perceptual and motor learning without conscious awareness. It involves the cerebellum and amygdala (see Refs.[34, 35]). The process of forming the memory goes through different stages. Recent memory is easily disrupted until the information is converted into a long-term memory. Then it is relatively stable. However, when time goes on, the stored informationm and the capacity to retrieve information gradually diminishes. The memory process undergoes a continuous change with time. For different types of learning one knows that the memory is not localized in just one particular place in the brain. For example, any of three visual pathways can sustain conditioning of heart rate response in pigeons.

It is known that the mechanism of learning on the cellular level involves change in the synapses. Implicit learning leads to changes in the effectiveness of the synaptic transmission. Establishing long-term memory requires the synthesis of new proteins and the growth of new synaptic connections. The storage of explicit memory in mammals uses long-term potentiation (LTP) in the hippocampus. Comparing the storage of memory in an electronic computer with that of the brain, one finds a big difference. In a computer information is stored locally with precise addresses. In the brain information is stored non-locally, it is distributed over a large number of synapses and eventually even several parts of the brain.

10.3 Learning in neural networks

Learning in neural networks can be done in basically two ways: supervised learning and unsupervised learning. The training of neural networks is most efficient, when learning is supervised. There are many network models using all kinds of variants of supervised learning rules, like feed-forward networks (perceptrons), recurrent networks allowing for connections and information flow in forward and backward direction, or Boltzmann machines (for an overview see [36]). However, in neurobiology, a learning process without supervision is more realistic. There is no teacher. This requires some self-organisation of neurons and connections. An learning rule realized in neurons has been proposed in 1949 by Donald Hebb [37]. He formulated what is known today as Hebb's learning rule: "When an axon of a cell A is near enough to excite cell B or repeatedly or persistently takes part in firing it, some growth process or metabolic change takes place in one or both cells such that A's efficiency, as one of the cells firing B, is increased". Unsupervised Hebbian learning networks have been widely applied to model the visual cortex. The formation of orientational columns in the visual field in young cats has been studied by von der Malsburg [38] and others. Linsker [39] proposed a model of self-organisation of the visual system which does not require structured input, i.e. he used random noise as input of the first layer. It is a muli-layer (modified) Hebbian learning network. As a result he observed in layer three the formation of center-surround cells (maximal response to bright spot in center of receptive field), other cells showed a 'Mexican hat" covariance function (nearby units were positively correlated, while distant units had a negative correlation). In other layers he observed synaptic weights deviating from circular symmetry, although the system had a symmetric architecture. It is remarkable that all those features emerged from random noisy input.

10.4 Neural network model of associative memory

The associative memory problem is considerd as an example where neural network models have given some insight into neuroscience. The Hopfield model [40] is considered as a kind of standard model to study this problem. The Hopfield model is given by a Hamilton function of configurations of patterns (composed of bits) plus an up-date rule for the evolution of the system. The Hamilton function is very similar to a Hamilton function used in condensed matter physics to describe ferro (or anti-ferro) magnetism. Because the working of memory in a mammalian brain involves a very large number of neurons, one has to consider any associative memory model as a many-body problem. In physics, one often encounters many-body systems. E.g., magnetism, conduction of electrons in metals, or superconductivity are many-body problems from condensed matter physics. Thus physicists have developed techniques to mathematically handle and eventually solve (at least approximately) such problems. One of those methods is to make use of statistical mechanics. In this case one assumes that the system is in a state of statistical equilibrium. This is a very strong assumption, which tells us a rule for the probability that the system can be found in state (configuration) for any given energy. It is the so-called Boltzmann-Gibbs distribution law. From the point of view of experimental neurobiology, this assumption is a gross oversimplification, if not to say, it is simply wrong. Nevertheless, it has turned out to be very useful for the purpose to solve the Hopfield associative memory model. This has been achieved by Amit, Gutfreund and Sompolinsky [41, 42]. The Hopfield model is mathematically equivalent to a model which describes a spin glass, the Sherrington-Kirkpatrick model. Its Hamiltonian

$$H = -\sum_{i<j} J_{ij}\sigma_i\sigma_j \qquad (26)$$

is similar to the Ising model of spins σ_i. However, the coupling between spins, J_{ij} are considered as random quenched variables, independent for any pair i,j and obeying a Gaussian distribution,

$$P[J_{ij}] = \prod_{i<j} \sqrt{N/2\pi}\exp[-J_{ij}^2 N/2] \ . \qquad (27)$$

Now randomness enters here two ways. First, the Hamiltonian has random coupling, which is a characteristic property of spin glasses, creating disorder. Second, the free energy function

$$F = -T\log Z \ , \qquad (28)$$

with Z being the partition function, is a function with a multiple valley structure. The local minima of this function can be interpreted as stored patterns of memory. The problem is to find those local minima. Now noise can be used as a helpful tool to solve this problem. A solution algorithm is simulated annealing, a Monte Carlo method, which works by starting from a high temperature, then gradually reducing the temperature allowing to arrive at the absolute minimum of the free energy. The idea of this algorithm is quite similar to the mechanism of stochastic resonance.

10.5 When lost in a foreign city how to find the railway station?

Suppose a tourist visits a foreign city. He has lost his map. He does not speak the local language. He can not ask anyone for help. He needs to find the railway station to catch the train. He needs to carry out a search and he needs a strategy. There are several strategies of searching. For example, he could mark down all streets he encounters, walk them along to the end. If he is unsuccessful, he would start with another street, which starts next to his present position. Alternatively, he could first walk down all streets going in south-north direction then all streets going in west-east direction. Or he could go in a kind of circles around a point, which he considers as a center (e.g the most crowded place he had encountered). He also could make a random walk, throwing a die at each corner, to choose the new direction. Which strategy would be most successful? Learning means to (a) try out different searching strategies, (b) memorize them, and (c) evaluate them by ranking their success. The next time being confronted with a similar problem, like e.g., searching a source of drinking water when lost in wilderness, he would recall his learned search strategy and in particular which were successful and which were not and act consequently.

Possible search strategies are: (i) Search without guidance. One can do a random walk and search or search in a walk following a regular pattern. The success will depend on parameters like step sizes ΔL_{step}, the number of targets N_{target}, the distribution of targets. Also it will depend if one uses a finite target volume ΔL_{target} (i.e., whenever the searcher has approached the target within a distance $d \leq \Delta L_{target}$, it means he has found the target) and also on the dimension of space (see Fig.[8]). (ii) Search with guidance. For instance, a bird of prey recognizes its target from some distance (which may be quite large). Insects are guided by the fragrance of flowers or insects by pheromones.

For instance if the target can be expressed as a local minimum of some potential or cost function, this represents a standard problem in mathematical optimisation theory. There are standard algorithms, like the steepest descent method, the conjugate gradient method, the method of simulated annealing or variants of genetic algorithms. In any case, searches with randomness/noise involved will in general be much more efficient (see sect. 12).

11 Small World and Scale Free Networks

The working of the brain it is not only determined by the number of neurons, but also - among other factors - by the "wiring" i.e. the connections of neurons. Recently, a new type of network architecture has been focused on - the so called "small world" networks [43] and also a variant the "scale-free" networks [44]. The small world networks are characterized by high local clustering. This means if node A is linked to node B and A is also linked to node C, then there is a high probability that B is linked to C. An equally important property is that there are short links to distant nodes. One can visualize such network architecture by thinking of nodes on a rectangular grid. Each node on a grid point is linked to his next-neighbor node on the grid (clustering). In addition there are a few links connecting pairs of nodes quite distant on the grid (short connections). The network architecture is somewhere between regular and random. The scale-free networks are characterized by many nodes having few links, some nodes having more links and few nodes having many links, the distribution being given by a power law. Those networks have been shown to explain the Milgram letter experiment. Such network architectures have been identified to occur in the internet and world wide web, in the distribution of powerlines in the US, in the biology of cellular and metabolic networks (of the nematode worm C. elegans, thoroughly studied in genetics). As an example from neuroscience, also the neural net of C. elegans follows such network architecture.

What has all this to do with probability and neuroscience? The clue is that this network has some random connections which apparently makes it very efficient. In other words, a purely regular network or a purely random network would be less efficient. This shows up in the fact that small world networks minimize the search time of addresses of nodes. Apparently, at least for the nervous system of C. elegans, small-world is the optimal architecture. Also Hodgekin-Huxley neurons have been investigated in computer simulations. It turned out that such neurons work optimal i.e. produce coherent oscillations and a fast system response, when they are linked in a small world topology [29]. This architecture has been explored also in computer simulations of learning. Finally, the small world network was found to be very efficient for the associative memory [45].

12 Algorithms

In mathematics it occurs often that deterministic problems can be solved efficiently via probabilistic/stochastic methods. As a first example consider integrals. In particular, consider integrals where the boundary of the integration domain is not a smooth surface, but has an irregular (crinkly) shape. In such cases a Monte Carlo computation using random nodes is often more efficient than using integration with nodes adapted to the geometry. Second, and more importantly, the stochastic computation of integrals is very effective (and in many cases the only way) to compute integrals in high dimensions. By a rule of thumb, when the dimension of the integration domain is larger than $d = 10$, then a Monte Carlo evaluation of the integral is more effecient than an evaluation using fixed node rules [46]. In physics it is a standard method to compute path integrals (where the integration domain has a dimension d in the order of a few thousand) via the Monte Carlo with importance sampling. A widely used algoritm is the Metropolis algorithm [47].

As a second example consider search an optimisation problems, like finding the ground state of a Hamiltonian in a high-dimensional configuration space, or of a highly disordered system like a spin-glass, or finding the shortest path in the travelling salesman problem. For such problems, variants of two methods are widely used and quite successful. One is the method of simulated annealing, a variant of the Monte Carlo Metropolis method. The other is the use of genetic algorithms. Both explore and search the configuration space using random numbers. For a comparison between both methods see [48].

As a third example consider deterministic differential equations, which can be expressed in terms of and solved by a stochastic process. That means, one can obtain the solution of a deterministic differential equation from a solution of a stochastic process. As a simple example think of the diffusion or heat equation and its solution in terms of a Monte Carlo simulation (Gaussian process, Brownian motion). It has been shown by Courant, Friedrich and Levy [49] that the solution of certain differential equations is equivalent to a random walk. As example consider the diffusion equation in d dimensions,

$$\Delta P(x, t; x_0, t_0) - \frac{1}{D}\frac{\partial}{\partial t} P(x, t, x_0, t_0) = 0, \tag{29}$$

imposing the initial condition

$$\lim_{t \to t_0} P(x, t; x_0, t_0) = \delta(x - x_0). \tag{30}$$

D is the diffusion coefficient. It is well known (for a proof see Ref.[50]) that the solution P can be obtained from a Gaussian random walk on a spatial regular lattice (lattice spacing a_s), where the time progesses also in discrete units (a_t). The solution of the differential equation is obtained from the solution of the random walk in the limit

$$a_s \to 0 \quad a_t \to 0, \quad \frac{a_s^2}{a_t} = 2d \ . \tag{31}$$

What have stochastic algorithms to do with neuroscience? The neuron performs tasks similar to a processor in a computer. As an example, the brain is very good at pattern recognition - a baby very early recognizes its mother. Pattern recognition is a cognitive task which can be formulated as a neural computation algorithm. Another example is the creation of an action potential. The neuron, according to the integrate-and-fire model works like a cash register with a summing device. It sums the action potentials incoming via dendrites from neighbor neurons. As a last example, recall that the synaptic neuro transmitter diffusion process is mathematically equivalent to the solution of some differential equation.

Given the fact that the neuron executes mathematical algorithms, like integration or solving differential equations, it is quite likely an advantage for the neuron to use algorithms based on stochastic methods (like on a computer with electronic processors). Note, however, that the neuron is not a digital computer, but rather an analogue one. So the algorithm is directly linked to and manifested in the architecture of a neuron. The conclusion, however, is the same: The stochastic behavior of a neuron is an algorithmic advantage in performing its tasks. This advantage may show up in faster execution of tasks or in the ability to perform several operations in parallel.

13 Conclusion

In Quantum mechanics, the concept of probability is fundamental. It is related to the existence of stable atoms, hence macromolecules and organic life. In neuroscience there is a vast number of viewpoints, where noise and stochastic behavior is beneficial. It is tempting to speculate how a brain relying only on a deterministic mode would look like. Quite likely it would not work.

Acknowledgments

This work is dedicated to Prof. M. El Naschie on the occasion of his 60th birthday. H.K. has been supported by NSERC Canada. H.K. is grateful for the kind hospitality at the SALK Institute, where part of this work has been done. H.K. is grateful for discussions with Prof. Francis Crick, SALK Institute, Prof. Terrence Sejnowski, SALK Institute, and Prof. Christof Koch, CalTech.

References

[1] J.K. Webb, M.T. Murphy, V.V. Flambaum, V.A. Dzuba, J.D. Barrow, C.W. Churchill, J.X. Prochaska, A.M. Wolfe, Phys. Rev. Lett. 87(2001)091301;
P.P. Avelino, C.J.A.P. Martins, G. Rocha, P. Viana, Phys. Rev. D62(2000)123508;
R.A. Battye, R. Crittenden, J. Weller, Phys. Rev. D63(2001)043505.

[2] R. Penrose, Shadows of the Mind, Oxford Univ. Press, Oxford (1994).

[3] E.O. Wilson, The Insect Societies, Harvard Univ. Press, Cambridge, MA (1971).

[4] G.J. Sussman, J. Wisdom, Science 241(1988)433.

[5] J. Laskar, Nature 338(1989)237.

[6] N. Murray, M. Holman, Science 283(1999)1877.

[7] A. Messiah, Quantum Mechanics, North Holland, Amsterdam (1969), Vol.I, p.129.

[8] H. Kröger, S. Lantagne, K.J.M. Moriarty, B. Plache, Phys. Lett. A199(1995)299.

[9] H. Kröger, Phys. Reports 323(2000)81.

[10] L.F. Abbot, M.B. Wise, Am. J. Phys. 49(1981)37.

[11] R.P. Feynman and A.R. Hibbs, *Quantum Mechanics and Path Integrals*, McGraw-Hill, New York (1965).

[12] D. Ferster, Science 273(1996)1812.

[13] D.J. Tolhurst, A.F. Dean, Vision Res. 1983(1983)775.

[14] F. Rattay, Chaos, Solitons & Fractals 11(2000)1875.

[15] P. Bak, C. Tang, K. Wiesenfeld, Phys. Rev. Lett. 59(1987)381;
P. Bak, Computers in Physics July/Aug 1991, 420;
S.K. Grumbacher, K.M. McEwen, D.A. Halverson, D.T. Jacobs, J. Lindner, Am. J. Phys. 61(1993)329.

[16] G. Nicholis, in *Physics of far-from-equilibrium systems and self-organisation*, in The New Physics, ed. P. Davies, Cambridge University Press, Cambridge (1989).

[17] P.J.E. Peebles, Principles of Physical Cosmology, Princeton Univ. Press, Princeton, New Jersey (1993).

[18] R. Benzi, S. Sutera, A. Vulplani, J. Phys. A14, L453 (1981).

[19] F. Moss, K. Wiesenfeld, Scientific American, Aug. 1995, p.66.

[20] K. Wiesenfeld, F. Moss, Nature 373(1995)33.

[21] L. Grammaitoni, P. Hänzi, P. Jung, F. Marchesoni, Rev. Mod. Phys. 70(1998)223.

[22] A. Longtin, J. Stat. Phys 70(1993)309.

[23] J.K. Douglass, L. Wilkens, E. Pantazelou, F. Moss, Nature 365(1993)337.

[24] J. Collins, T. Imhoff, P. Grigg, Nature 383(1996)770.

[25] D. Nozaki, J.M. Douglas, P. Grigg, J.J. Collins, Phys. Rev. Lett. 82(1999)2402.

[26] J.E. Levin, J.P. Miller, Nature 380(1996)165.

[27] L.C. Gebeshuber, A.V. Holden, D. Petrachi, eds., Stochastic Resonance in Biological Systems, Special Issue, Chaos, Solitons & Fractals Vol.11, No.12 (2000).

[28] A.J. Noest, Phys. Rev. Lett. 63(1989)1739.

[29] L.F. Lago-Fernandez, R. Huerta, F. Corbacho, J.A. Sigüenza, Phys. Rev. Lett. 84(2000)2758.

[30] E.R. Kandel, Kleine Verbände von Nervenzellen, in Gehirn und Nervensystem, Spektrum der Wissenschaften (German Edition of Scientific American), Heidelberg (1988).

[31] T.V.P. Bliss and M.A. Lynch, in P.W. Landfield and S. Deadwyler, eds., Long-Term Potentiation: From Biophysics to Behavior, A.R. Liss, New York (1988), p.3.

[32] J.H. Byrne and S.G. Schultz, An Introduction to Membrane Transport and Bioelectricity, Raven Press, New York (1994).

[33] T.J. O'Dell, R.D. Hawkins, E.R. Kandel and O. Arancio, Proc. Nat. Acad. Sci. (USA) 88(1991)11285-9;
E.M. Schuman and D.V. Madison, Science 254(1991)1503-6.

[34] E.R. Kandel, J.H. Schwartz and T.M. Jessell, Essentials of Neural Science and Behavior, Appleton & Lange, Norwalk (1995).

[35] L.R. Squire, Memory and Brain, Oxford Univ. Press, New York (1987).

[36] J. Hertz, A. Krogh, R.G. Palmer, Introduction to the Theory of Neural Computation, Addison-Wesley Publ., Redwood City, CA (1991).

[37] D.O. Hebb, Organisation of Behavior, Wiley, New York (1949), p.62.

[38] C. von der Malsburg, Kybernetik 14(1973)85;
C. von der Malsburg, J.D. Cowan, Biological Cybernetics 45(1982)49.

[39] R. Linsker, Computer, March 1988, p.105.

[40] J.J. Hopfield, Proc. Nat. Acad. Sci. (USA) 79(1982)2554-2558;
see also J. Hertz, A. Krogh and R.G. Palmer, Introduction to the Theory of Neural Computation, Addison Wesley, Reading (1991).

[41] D. Amit, H. Gutfreund and H. Sompolinsky, Phys. Rev. A32(1985)1007-1018; Phys. Rev. Lett. 55(1985)1530-1533; Ann. of Phys. 173(1987)30-67; Phys. Rev. A35(1987)2293-2303.

[42] D. Amit, Modelling Brain Function, Cambridge Univ. Press, Cambridge (1988).

[43] D.J. Watts, S.H. Strogatz, Nature 393(1998)440;
D.J. Watts, Small Worlds, Princeton Univ. Press, Princeton, NJ (1999);
J.M. Kleinberg, Nature 406(2000)845;
S.H. Strogatz, Nature 410(2001)268;
D.J. Watts, P.S. Dodds, M.E.J. Newman, Science 296(2002)1302.

[44] A.L. Barabasi, R. Albert, Science 286(1999)509;
R. Albert, H. Jeong, A.L. Barabasi, Nature 401(1999)130;
R. Albert, H. Jeong, A.L. Barabasi, Nature 406(2000)378;
H. Jeong, B. Tombor, R. Albert, Z.N. Oltavi, A.L. Barabasi, Nature 407(2000)651;
A.L. Barabasi, Linked - the New Sience of Networks, Perseus Publishing, Cambridge, MA (2002).

[45] J.W. Bohland, A.A. Minai, Neorocumputing 38-40(2001)489.

[46] F. James, Rep. Prog. Phys. 43(1980)1145.

[47] N. Metropolis, A. Rosenbluth, M. Rosenbluth, A. Teller and E. Teller, J. Chem. Phys. 21(1953)1087.

[48] D.R. Thompson, G.L. Bilbro, IEEE Comm. Lett. Vol.4, No.8(2000)267.

[49] R. Courant, K.O. Friedrichs, H. Levy, Math. Ann. 100(1928)32.

[50] C. Itzykson and J.M. Drouffe, Statistical Field Theory, Cambridge Univ. Press, Cambridge (1989), Vol.I.

Ramifications of Non Commutative Spacetime

by B.G. Sidharth

Abstract

We review Cantorian and Non Commutative Spacetime, work which has occupied El Naschie in the past several years. These concepts are now the subject of intense research, thanks to Quantum Gravity, Quantum Super String Theory and a few other approaches. It now appears that we are on the verge of a breakthrough in finding solutions to longstanding problems, like the unification of gravitation with other fundamental interactions and the question of the mass spectrum, another recent area of El Naschie's work.

1 Introduction

It is now a cliche that the two great intellectual pillars of the twentieth century, viz., General Relativity or, more generally, Gravitation and Quantum Mechanics have stood apart, stubbornly defying attempts at unifying them. As Wheeler [1] noted, the problem finally boils down to the introduction of the Quantum Mechanical concept of spin half into Classical Theory and the classical concept of curvature into Quantum Theory. Curiously enough the two pillars of General Relativity and Quantum Theory stand on a common ground: Together they use the concept of a differentiable space time manifold, be it the Reimannian space time of General Relativity or the Minkowski space time of Relativistic Quantum Theory or Quantum Field Theory. However more recent work be it in Quantum Gravity or in Quantum Super String Theory, has hinted at a minimum space time cut off [2, 3, 4]. This alters the age cold concept of a smooth space time. In recent years there has been quite some work by different authors such as Ord, Nottale, El Naschie, the author and others in this relatively new field of non differentiable space time [5, 6, 7, 8, 9, 10, 11] and several references therein. One of the very fruitful concepts put forward in these pathbreaking efforts has been El Naschie's concept of the Cantorian space time. Atlast some schoalrs are beginning to realize the fractal nature of space time. Once we break out of the smooth spacetime mindset, many exciting possibilities open up, including a unification of gravitation with other fundamental interactions, and also the possibility of solving the elusive problem of the mass spectrum, as has been deduced by El Naschie for example [38].

We will briefly survey some of these efforts and indicate how the solution of longstanding problems are now within sight.

catalog called "space time" ...."[1]. This approximate classical space time is quasi- changeless or stationary and time is reversible, as indeed is evident from the equations of motion both in Classical Physics and Quantum Physics. But when we go beyond this approximation, to the stochastic description at the Planck scale [23] time is no longer reversible. We make the leap from the age old concept of "being" to the concept of "becoming" at least at this scale.

One of the criticisms put forward against the Planck Scale Phenomena is that these effects are beyond experimental verification for a long time to come. But let us analyse this further. The Compton scale which we encounter in the physical world has an underpinning of some $n = 10^{40}$ transient Planck particles. However Planck scale phenomena are moderated, and we have, as in a diffusion process,

$$l = \sqrt{n} l_P \qquad (3)$$

$$m = m_P/\sqrt{n} \qquad (4)$$

where l and m are the Compton wavelength and mass of a typical elementary particle and l_P and m_P are the Planck length and the Planck mass. An equation identical to (3) holds for the Compton time also.

(3) and (4) are not mere numerical accidents - they can be deduced as mentioned from a diffusion process [23]. The $\sqrt{n}$ in the equations is indicative of a Brownian process. For example, in a random walk of n steps, each of length l, the total distance covered would be of the order of $\sqrt{n} l$.
Infact we can go one step further. Remembering that there are a total of $N = 10^{80}$ elementary particles, the entire universe shows up as $n \times N = 10^{120}$ Planck scale oscillators. Using the fact that the rth energy level for the Harmonic oscillators is given by $\sqrt{r}\hbar\omega$ for large r, [25] it follows that the total energy of these Planck scale oscillators would be $\sqrt{nN} m_P c^2$, which correctly gives the mass of the universe itself. That is, the universe is a normal mode of these Planck scale oscillators.

We have already remarked that the Quantum commutators are present in (1), if we neglect terms of the order of a^2. In particular, taking a to be the Compton scale, it has been shown that we can recover from (1) and (2) the Dirac equation itself[11]. This again is not surprising because the non commutativity in (1) and (2) can be shown to represent spin even from the classical viewpoint [26]. Interestingly, at the Compton scale it has been shown that the Quantum coordinates coincide with the complex coordinates of a classical Kerr-Newman Black Hole with radius of the order of the Compton wavelength. Indeed it has been known for a long time that the classical Kerr-Newman metric reproduces the field of the electron including the purely

Quantum Mechanical gyro magnetic ratio $g = 2$. What has been inexplicable is the fact that there is a naked singularity or equivalently complex coordinates. This infact is a direct consequence of the non commutativity [11]. In other words once the non commutative or fuzzy nature of spacetime is taken into consideration, corresponding to averages over zitterbewegung in Realtivistic Quantum Theory, the naked singularity disappears and the electron can be represented by the Kerr-Newman metric. These conclusions have since been confirmed by Nottale [27]. Already this Kerr-Newman characterization of the Quantum Mechanical electron points to the long sought after linkage between gravitation and electromagnetism [28]. It can be shown formally that this is so [29, 20, 22]. Infact arising from (1) there is the covariant derivative

$$\partial_\mu \to \partial_\mu - \Gamma^\sigma_{\mu\sigma} \tag{5}$$

The second term on the right of (5) represents the electromagnetic potential, and surprisingly coincides with the original Weyl formulation of electromagnetism.

Surprising as it may seem there is a cosmological scheme which follows from these considerations. It can be understood on the basis of the fact that in the minimum time interval τ at the Compton scale, $\sqrt{N}$ particles are created fluctuationally from the Quantum vacuum, where N is the number of particles present at that epoch [30, 31, 32]. This cosmology successfully predicted dark energy and an accelerating ever expanding universe since confirmed by observation [33, 34].

All this can also be shown to explain some hitherto inexplicable coincidences noted by Weyl, Eddington, Dirac and Weinberg. These are relations like

$$R = \sqrt{N} l \tag{6}$$

$$\frac{e^2}{Gm^2} = \sqrt{N} \tag{7}$$

$$m = \left(\frac{\hbar^2 H}{Gc}\right)^{1/3} \tag{8}$$

and others. R is the radius of the universe, e is the electron charge, G the universal constant of gravitation, c the velocity of light, $\hbar$ the reduced Planck constant and H the Hubble constant. It is easy and even unscientific to dismiss such equations as accidents. Dirac himself realised that equations (6) and (7) for example could have a cosmological significance [35]. There was an inconsistency in his otherwise beautiful cosmology. More recently it has been shown by the author that these coincidental equations (8) can be deduced, on the basis of the cosmology mentioned above. It may be mentioned that

Weinberg had termed the equation (8) as being mysterious because it relates a large scale parameter like H, the Hubble constant to microphysical constants [36]. This points to a Machian or "co-related" universe which is not surprising because the universe as we have just seen, is a normal mode of Planck oscillators.

In this consistent scheme, the universe is created out of a sub stratum Quantum vaccuum or dark energy, in a phase transition at the Planck scale [37]. Herein are the very first seeds for all the complex structures of the universe.

3 A Mass Spectrum

We can extend these considerations to generate a mass spectrum, a problem that has fascinated El Naschie and for which his model gives a solution [38]. We now use the model of three oscillators (typically for the three quarks), discussed in detail in references [39, 40, 11, 41]. We use the fact that for such an oscillator (resembling a triatomic molecule) [42] the frequencies are given by

$$\omega = \sqrt{k/m}, 2\sqrt{k/m} \qquad (9)$$

for one and two such oscillators where $\hbar\omega = mc^2$, m being the mass.

For reasons discussed in detail in the references, we take the higher frequency or mass to represent the pion m_π. In this connection, it may be mentioned that the π^0 meson has been shown to be a bound state of an electron and positron in the discrete spacetime theory [11], borne out by its decay mode. This was a starting point for El Naschie in his mass spectrum model. Then from (9) we derive the mass spectrum from the oscillator energy levels viz.,

$$m \approx (n + \frac{1}{2})m_\tau, m \approx (2n+1)m_\pi, n = 0, 1, 2, \cdots \qquad (10)$$

It is remarkable that the formula (10) generates a whole range of some 50 mesons and baryons including the well known particles from the particle data group tables (Cf.refs.[43]), starting from the K meson ($480 meV$ approximately) through several other mesons including the $2\chi, D, f, \rho, \Theta$ and so on as also the very massive $\gamma(1s), \chi 1p, \gamma 2s, \gamma 3s, \gamma 4s, \gamma 10860, B, J/(\psi 1s)$ and so on, going right up to values of n near 80 giving the heaviest of the particles and clusters of masses nearby. Even in the approximation (10) where several degrees of freedom and other details have been excluded, the agreement is within a few percent of the actual values. Interestingly the latest particle $D_5(2317)$ [44] also follows from (10), being apprxomately $17m_\pi$. Incidentally El Naschie's mass spectrum can also start with the model of spring connected oscillators [38].

We can push the above ideas further, using the Coulumbic part of the interquark potential to exhibit an array of two or three or more pions as ascillators (like a tri-atomic molecule) with different anergies and show that the various elementary particles correspond to the various energy levels [B.G. Sidharth, Physics/0306010 and B.G. Sidharth, Physics/0309037]. The simple formula that we get is, a slightly generalized version of (10) viz.,

$$\text{Elementary Particle mass} \simeq m(n + \frac{1}{2})m_\pi \qquad (11)$$

where m and n are positive integers and m_π is the pion mass. It is easy to verify that (11) gives the mass of all known elementary particles with a maximum error of about three percent. Moreover the pentaquark which was discovered subsequently also agrees with (11).

References

[1] C.W. Misner, K.S. Thorne and J.A. Wheeler, "Gravitatioin", W.H. Freeman, San Francisco, 1973, pp.448ff.

[2] D. Amati, in Sakharov Memorial Lectures, Eds. L.V. Kaddysh, N.Y. Feinberg, Nova Science, New York, 1992, p.455ff.

[3] L.J. Garay, Int.J.Mod.Phys. A., 1995, 10(2), p.145-65.

[4] B.G. Sidharth, Chaos, Solitons and Fractals, 15, 2003, p.593-595.

[5] G.N. Ord, Int.J.Th.Phys., Vol.35, No.2, 1996, p.263-266.

[6] G.N. Ord, Int.J.Th.Phys., Vol.31, No.7, 1992, p.1177-1195.

[7] L. Nottale, "Fractal Space-Time and Microphysics: Towards a Theory of Scale Relativity", World Scientific, Singapore, 1993, p.312.

[8] M.S. El Naschie, Chaos, Solitons and Fractals, 7(4), 1996, pp.499-518.

[9] M.S. El Naschie, Chaos, Solitons and Fractals, 10(11), 1999, pp.1813-1819.

[10] M.S. El Naschie, Int.J.Th.Phys., Vol.37, No.12, 1998, p.2935ff.

[11] B.G. Sidharth, "Chaotic Universe: From the Planck to the Hubble Scale", Nova Science Publishers, Inc., New York, 2001 (And several references therein).

[12] P.A.M. Dirac, "The Principles of Quantum Mechanics", Clarendon Press, Oxford, 1958, pp.4ff, pp.253ff.

[13] F. Rohrlich, "Classical Charged Particles", Addison-Wesley, Reading, Mass., 1965.

[14] B.G. Sidharth, in Instantaneous Action at a Distance in Modern Physics: "Pro and Contra", Eds., A.E. Chubykalo et.al., Nova Science Publishing, New York, 1999.

[15] Gerard 't Hooft in "Frontiers of Fundamental Physics 4", Kluwer Academic/Plenum Publishers, New York, 2000, p.1-12 ff.

[16] J.R. Klauder, "Bosons Without Bosons", in Quantum Theory and The Structures of Time and Space, Vol.3, Eds. L. Castell, C.F. Van Weiizsecker, Carl Hanser Verlag, Munchen, 1979.

[17] M. Sachs in "Directions in Microphysics", Fondation de Broglie, Paris, 1993, pp.393ff (For Dirac quote).

[18] H.S. Snyder., Physical Review, Vol.71, No. 1, January 1947, p.38-41.

[19] H.S. Snyder., Physical Review, Vol. 72, No.1, July 1 1947, p.68-71.

[20] B.G. Sidharth, Annales de la Foundation Louis de Broglie, Vol. 27, No.2, 2002, pp.333-342.

[21] J. Polchiroski, Rev.Mod.Phys., 65(4), 1996, pp.1245ff.

[22] B.G. Sidharth, Nuovo CimentoB, **117B** (6), 2002, p.703.

[23] B.G. Sidharth, Found.Phys.Lett., 15 (6), 2002, pp.577-583.

[24] B.G. Sidharth, "Space-time" in Concise Encylopaedia of Supersymmetry and Noncommutative Structures", Eds. J. Bagger, S. Dupliz and W. Siegel, Kluwer, Dordrecht, 2001.

[25] B.G. Sidharth, "Fuzzy Space Time, Quantum Geometry and Cosmology", to appear in Proceedings of Frontiers of Fundamental Physics Vol.5., Universities Press (Orient Longman).

[26] S. Zakrewski, "Quantization, Coherent States and Complex Structures", Ed. J.P. Antoine et al, Plenum Press, New York, 1995, p.249ff.

[27] L. Nottale, Chaos, Solitons and Fractals, 12(9), July 2001, pp.1577ff.

[28] B.G. Sidharth, Gravitation and Cosmology, 4(2), (14), 1998, p.158ff.

[29] B.G. Sidharth, Nuovo Cimento, **116B** (6), 2001, p.735ff.

[30] B.G. Sidharth, Int.J.Mod.Phys.A., 13 (15), 1998, p.2599ff.

[31] BG. Sidharth, Int.J.Th.Phys., 37 (4), 1998, 1307-1312.

[32] B.G. Sidharth, Chaos, Solitons and Fractals, 16(4), May 2003, pp.613-620.

[33] S. Perlmutter, et al., Nature, Vol.391, 1 January 1998, p.51-59.

[34] R.P. Kirshner, Proc. Natl. Acad. Sci. USA, Vol.96, April 1999, pp.4224-4227.

[35] J.V. Narlikar, "Introduction to Cosmology", Cambridge University Press, Cambridge, 1993, pp.237ff.

[36] S. Weinberg, "Gravitation and Cosmology", John Wiley & Sons, New York, 1972, pp.619ff.

[37] B.G. Sidharth, Chaos, Solitons and Fractals, 18, 2003, 197-201.

[38] M.S. El Naschie, Chaos, Solitons and Fractals, 14, 2002, p.649-668.

[39] B.G. Sidharth, Mod.Phys.Lett.A., Vol.14, No.5, 1999, p.387-389.

[40] B.G. Sidharth, Mod.Phys.Lett.A., Vol.12, No.32, 1997, p.2469-2471.

[41] Y. Yu Lobanov in "Frontiers of Fundamental Physics", Eds. B.G. Sidharth and A. Burinskii, Universities Press, 1999, p.135ff.

[42] D.L. Goodstein, "States of Matter", Dover Publications Inc., New York, 1985, p.160ff.

[43] K. Hagiwara, et al., (Particle Data Group), Phys.Rev.B., <u>66</u>, 010001.

[44] Nature, Report, 8 May, 2003.

Computational Universes

by Karl Svozil

Abstract

Suspicions that the world might be some sort of a machine or algorithm existing "in the mind" of some symbolic number cruncher have lingered from antiquity. Although popular at times, the most radical forms of this idea never reached mainstream. Modern developments in physics and computer science have lent support to the thesis, but empirical evidence is needed before it can begin to replace our contemporary world view.

1 Historical Notes

In a broad context, the development of rationalism, the enlightenment and science can be perceived as an awakening from the illusory world of the senses (*Maya* in Sanskrit); as a growing awareness that "facts" which once were perceived as self-evident turned out to be utterly wrong. Humanity once took it for granted that it was located at the epicentre of the Universe. A closer inspection revealed that there is no ground to claims of any preference in location: Earth is conveniently situated in a solar system of a remote part of our galaxy, which in turn is part of a group of galaxies and of the physical Universe as we perceive it today. People also trusted that their bodies are made-up of solid stuff. Later on they learned that, as their bodies consist of atomic and subatomic "point" particles, things only appear to be solidly filled, but in another perspective, space is "almost empty." Time turned out to be relative to the motion of observers, and single "particles" such as photons and neutrons, seemed to be at two or more spatial positions at once. On another issue, people previously thought that they have been created in a different way than other species. As it turned out, from a biological point of view, mankind evolved and spread just like locusts and everyone else around. This is corroborated not only by phylogenetic evidence, but by analysis of the very DNA code that constitutes the genetic heritage and blueprint of our ancestors and of all living beings. Indeed the DNA itself turns out to be a biochemical code running on cellular computers to the effect of creating, maintaining and reproducing the organism of which it is a part.

Further disillusionments may lie ahead. Consciousness is still an "undiscover'd country," and maybe it is just a manifestation of neuronal brain functions. Or, consciousness may be just the opposite: transcendental. Despite the achievements of Freud, certain dream phases are barely understood.

Artists have speculated that we are "fleshware" units inside of a simulation-computation-game that appears gigantic or even infinite to us. Who knows, we might have even paid for to live a life in the twenty-first century in a beyond fair. That is to say, we might be embedded in a literal "game" that we chose to pass the time. To make things more realistic, all memories of the our life in the beyond might have been erased from our immediate memories[1]. Maybe the "meaning" of our world is rather trivial; like the simulation of marketing measures for a beyond world[2]. As computers have begun to permeate our societies, it is no wonder that the "universe as a computer" metaphor for the physical Universe has attracted increasing attention. Perhaps some day our own technology could achieve such visions, and put it to our practical use[3].

In antiquity, Pythagoras (6th cent. B.C.) *"considered numbers as the essence and principle of all things, and attributed to them a real and distinct existence; so that, in his view, they were the elements out of which the universe was constructed"* (from Bulfinch [5]). Plato's (c. 427- c. 347 B.C.) emphasis in geometry, in particular his dictum *"God geometrizes"* [4] was interpreted by Gauss (1777-1855) as *"o theos arithmetizei,"* or *"God computes."* The vision of a clockwork universe is probably best characterized by the (probably apocryphal) story, that when Laplace was asked by Napoleon how God fitted into his secular system of *Mécanique Céleste*, he replied [6, p. 538], *"I have no need for that hypothesis"*[5].

In his famous lecture delivered before the International Congress of Mathematicians at Paris in 1900, Hilbert (1862-1943) enumerated twenty-three prob-

---

[1] This is mind-body dualism in a new form. For a concrete mind-brain interface model, see for instance Eccles' proposal [1]. In this view, what appears to us as the physical world is just a simulation-computation-game; and the mind(s) of the player(s) is (are) transcendental with respect to the characters in this emulation. Note the phrase *"we are the dead on vacation"* in Godard's film *Breathless*.

[2] This is the theme in Galouye's 1964 novel *Simulacron 3*; the novel stimulated Fassbinder's *Welt am Draht*, as well the recent movie *Thirteenth Floor*. Somewhat related scripts are those of *Total Recall* and *Matrix*. In *Contact*, Sagan mentions the "Zoo hypothesis" claiming that there is somebody (in this case aliens) watching us for ethnographic or other reasons. Philosophical speculations include Rene Descartes' *world-as-a-lucid-dream* vision [2, Meditation 1,9], and Putnam's *brain-in-a-vat* metaphor [3]; see also http://whatisthematrix.warnerbros.com.

[3] There is a possibly apocryphal story [4, p. 127] that, when asked by his Prime Minister Peel or by the Chancellor of the Exchequer Gladstone about the usefulness of his findings, Faraday responded, *"Why, sir, there is the probability that you will soon be able to tax it."*

[4] In *Convivialium disputationum, liber 8,2,* Plutarch stated, *"Plato said God geometrizes continually."*

[5] In his memoires written on St. Hélène, Napoleon states that he removed Laplace from office as Minister of the Interior [6, p. 536] after only six weeks *"because he brought the spirit of the infinitely small into the government."*

lems, among them the compatibility of the arithmetical axioms (# 2), the mathematical treatment of the axioms of physics (# 6), and the determination of the solvability of a diophantine equation (# 10)[6]. Gödel (1906-78), as well as Turing (1912-54) contributed towards the (negative) solution of # 2 & # 10. They pursued a formalization of mathematics by coding of axiomatic systems, either by the uniqueness of prime factorization or by their representation as (universal) computer programs[7].

For the first time in human history, we are able to articulate precisely what we mean when discussing computations. Turing's universal computer model is modelled after the syntax of everyday pencil and paper operations which children learn at school. The paper lines are unwound into a tape, and whatever rules there are for computing can be represented by the combination of tape, finite memory and simple read-write operations of the Turing machine.

The notion of universal computation is *robust* in the sense that any universal computer can emulate any other universal computer (regardless of efficiency and overhead), so that it does not really matter which one is actually implemented. In a sense, the entire class of universal computer counts as a single computer, because they are all equivalent with respect to algorithmic emulation of one another.

Robustness is a very important concept for the matter of computational universes, because it is not really important on which particular models or hardware these universes are implemented; they are all in the same equivalence class. Apart from the translation from one coding scheme to another, each one of them is equivalent to the entire class. So, when it comes to their generic properties, it is not really important whether automaton universes are modelled to be Cellular Automata, Turing Machines, colliding billiard balls [8], or biological substrates. All of this means that one is free to choose whatever computational model suits best the particular purpose one has in mind.

---

[6] http://babbage.clarku.edu/~djoyce/hilbert/problems.html

[7] In a postscript dated from June 3rd, 1964 [7, p. 369-370], Gödel's opinion is clearly expressed, "... due to A. M. Turing's work [[on the universal Turing machine]], a precise and unquestionably adequate definition of the general concept of formal system can now be given, the existence of undecidable arithmetical propositions and the non-demonstrability of the consistency of a system in the same system can now be proved rigorously for every consistent formal system containing a certain amount of finitary number theory. ... Turing's work gives an analysis of the concept of "mechanical procedure" (alias "algorithm" or "computation procedure" or "finite combinatorial procedure"). This concept is shown to be equivalent with that of a "Turing machine." A formal system can simply be defined to be any mechanical procedure for producing formulas, called provable formulas."

Gödel, Tarski, Turing and Chaitin, among others, revealed that, stated pointedly, mathematical "truth" extends formal "provability." Mathematics is incomplete, and there always will be true theorems about a particular formal system of axioms (sufficiently rich to contain arithmetic), such as consistency, which are not provable "from within" that system[8].

Wigner considered *"the unreasonable effectiveness of mathematics in the natural sciences"* [10], which is usually taken for granted but which, upon inspection, seems unfounded. One obvious solution to this bewilderment seems to be the Pythagorean assumption that numbers are the elements out of which the universe was constructed; and what appears to us as the laws of Nature are just mathematical theorems or computations. Notice that, whereas Gödel once and for all settled the question of a complete finite description of mathematics to the negative, the question of whether or not a finite mathematical treatment of the axioms of physics exists (Hilbert's problem # 6) remains open.

Another thread was opened by Edward Moore. Puzzled by the quantum mechanical feature of complementarity, Moore conceived a finite deterministic model of complementarity capable of being run on a computer [11, 12]. This formalization of complementarity, not in terms of Hilbert space quantum mechanics, but by constructive algebraic, even finitistic, means, may be perceived as the continuation of the Turing program to formalize the notion of "algorithm" or "computation" by conceptualizing it as a concrete machine model.

In another development, Von Neumann (preceded by Ulam [13]) constructed a two-dimensional cellular array of finite deterministic automata which are connected to their neighbors such that the state of each one of these automata is determined by the previous states of itself and of its neigbors [9]. He was able to show that such cellular automata (CA) could not only be in the robust class of universal computers, but that entities inside such arrays could reproduce themselves by holding their own descriptional code and the algorithmic means to construct identical copies of themselves.

---

[8]Gödel's own thoughts on the interpretation of his results are formulated very nicely in a reply to a letter by A. W. Burks, reprinted in [9, p. 55], *"I think the theorem of mine which von Neumann refers to is ... the fact that a complete epistemological description of a language A cannot be given in the same language A, because the concept of truth of sentences of A cannot be defined in A. It is this theorem which is the true reason for the existence of undecidable propositions in the formal systems containing arithmetic. I did not, however, formulate it explicitly in my paper of 1931 but only in my Princeton lectures of 1934. The same theorem was proved by Tarski in his paper on the concept of truth ..."*

Stimulated by Von Neumann's concept of CA, Konrad Zuse, the creator of one of the first general purpose digital computers, suggested to look into the idea that physical space itself might actually be such a "calculating space" ("Rechnender Raum") [14, 15, 16]. In this view, the physical objects exist as computational entities immersed in such a computational medium. Zuse became fascinated by the idea of going beyond quantum mechanics in discretizing physics [9], a vision he shared with the late Einstein [10] and many researchers, among others Fredkin, Toffoli, Margolus, and Wolfram.

Fredkin and Toffoli investigated reversible CA, in which the global temporal evolution can be inverted uniquely. That is, any CA configuration has a unique predecessor and a unique successor. Note that, if the evolution is a bijective map; i.e., is one-to-one for every single cell, then the global array is a reversible CA as well. (The converse need not be satisfied.) For a concise account [11] the reader is referred to the reviews by Toffoli & Margolus [19], Fredkin [20] and Wolfram [12]. In a reversible world, nothing is lost or gained; and all revelations are permutated back and forth. In this sense, the very concept of question and answer, of problem and solution, of past, present and future, and thus of a directed 'lapse of time," remains relative, subjective and conventional [23].

2   Intrinsic Randomness and Undecidability

Contemporary theoretical physics postulates at least three types of randomness: (i) the "chaotic" randomness residing in the initial conditions, which are

---

[9]Quantum theory just discretizes the number of quanta within a mode, yet the modes themselves are still continuous.

[10]In [17, p. 163], Einstein states, *"There are good reasons to assume that nature cannot be represented by a continuous field. From quantum theory it could be inferred with certainty that a finite system with finite energy can be completely described by a finite number of (quantum) numbers. This seems not in accordance with continuum theory and has to render trials to describe reality with purely algebraic means. However, nobody has any idea of how one can find the basis of such a theory."*

[11]Reversibility of CA should not be confused with Bennett's strategy to produce "reversible" calculations from irreversible ones by temporarily copying their intermediate results and permanently copying their final result, thereby setting the computing agent to its initial state, as well as retaining the result of the computation [18]. Any such operation must necessarily allow for copying; i.e., for a one-to-many evolution, which is clearly not meant here.

[12]Wolfram recently self-published a long-awaited and widely publicised book which, among other issues, deals with some of the rules categorized for one-dimensional automata [21] and their conceivable physical applications. It has been received ambivalently with reviews ranging from the author doing nice computer graphics to becoming the biggest physics guru of all times [22]. Wolfram has attracted a lot of attention for this subject; and it can only be hoped that the many claims made in this *opus* will not deter others.

assumed to be "drawn" (*via* the postulated axiom of choice) from a "continuum urn." Almost all elements of the continuum are nonrecursively enumerable and even random; i.e., algorithmically incompressible [24, 25, 26]; (ii) the random occurrence of individual quantum events such as a detector click; (iii) complementarity; i.e., the impossibility to measure two or more observables with arbitrary precision at once.

2.1 Computational Complementarity

As already mentioned, Moore [11] invented (parts of) finite automata theory to formalize and model physical complementarity. Research in this area became totally separated from its original physical perspective and developed into a beautiful algebraic theory of its own [27]. Finkelstein [28] rediscovered Moore's paper and coined the term "computational complementarity." Its concrete logico-algebraic structure has been investigated by the author in a series of papers with C. and E. Calude, Khoussainov, Lipponen, Schaller and others [29, 30, 31, 32], also in the context of reversible computation [33, 34]. Automaton logic turns out to be logically equivalent [35] to generalized urn models [36, 37], indicating that the associated logico-algebraic structure is more robust than could be assumed from those single model types alone.

Arguably, the simplest automaton model featuring complementarity is a finite (Mealy) automaton in which the sets contain three internal states $S = \{1, 2, 3\}$, three input symbols $I = \{1, 2, 3\}$, and two output symbols $O = \{0, 1\}$. Let, for $s \in S$, $i \in I$, the (irreversible "guessing") output function be $\lambda(s, i) = \delta_{si}$. The (irreversible) transition function just steers the automaton into a state corresponding to the input symbol; i.e., $t(s, i) = i$. The problem of finding an unknown initial state by analysis of experimental input–output sequences yields a partitioning of the internal states $\{\{1\}, \{2, 3\}\}$, $\{\{1, 3\}, \{2\}\}$, and $\{\{1, 2\}, \{3\}\}$, according to the input 1, 2, and 3, respectively. Every one of the partitions constitutes a Boolean algebra whose elements are comeasurable. The pasting of these three Boolean algebras along their common elements (in this case just $\emptyset$ and $\{1, 2, 3\}$) yields a nonclassical, nondistributive logical structure MO_3, which is also realized by the algebra of propositions associated with the electron spin state measurements along three different directions.

A systematic investigation shows that the logico-algebraic structures arising from computational complementarity are very similar to those encountered in quantum logics [32, Sec. 3.5.2]. In particular, any finite quantum (sub-)algebra can be represented as an automaton logic and thus can be modelled with a finite automaton. Clearly, infinite quantum structures, such as the continuous "Chinese lantern" lattices MO_c involved in electron spin state measurements in continuous directions, or quantum contextuality, cannot be modelled with a finite automaton.

Reversible finite automata have been introduced by the author [33, 32, 34] as Mealy automata whose input and output symbols are identical. Consider the Cartesian product $S \times I$ of the set of automaton states S with the set of input symbols I, arranged in vector form $SI = ((s_1, i_1), \ldots)$; as well as the Cartesian product $S \times O$ of the set of automaton states with the set of output symbols O, again arranged in vector form $SO = ((s_1, o_1), \ldots)$. The transition and output functions and thus the automaton computation can then be formalized by a matrix multiplication $SO = SI \cdot P$, where P is the matrix associated with the combined transition and output function $P : SI \to SO$. Reversibility implies that these matrices P are permutation matrices (i.e., every row and every column contains exactly one entry "1," all other entries are zeroes).

The most general probabilistic state of all reversible Mealy automata associated with a particular matrix "dimension" can be represented as the weighted convex sum over all permutation matrices of this dimension. The result is a doubly stochastic matrix (i.e., the sum of the real components of every row and column adds up to one). Formally, let $\psi : SI \times SO \to [0, 1]$ be the transition probability. The convex sum of all transition probabilities is one; i.e., $\sum_{SI,SO} \psi(SI, SO) = 1$.

A modification of this model, according to Fortnov [38, 39], captures the class **BQP**, the class of efficiently quantum computable problems. The modification is twofold: first, the weighted sum over all permutation matrices contains coefficients ψ, called "probability amplitudes," which take on arbitrary rational values including *negative* values. Secondly, in order for the "quantum" probabilities to be positive, the probability amplitudes ψ have to be squared. These two modifications—negativity and square values—mark a demarcation line between quantum and classical computation.

Although computational complementarity will not be reviewed any further here, it should be mentioned that Moore conceptualized input/output experiments on finite automata, making a distinction between "intrinsic" cases where only one automaton is available and those in which an arbitrary number of identical copies are accessible. In the latter case, there is no complementarity, because after any experiment it is always possible to dispose of the used automaton and get a fresh automaton copy for further experiment(s).

The intrinsic, embedded, case is the one experienced in physics, because the observer cannot escape and always is part of the ("Cartesian prison" [2, Meditation 5,15]) system. Due to restrictions in copying and cloning, it is not possible, for instance, to obtain an identical copy of a single photon or electron in a nonclassical state. And only in the single automaton case there is a chance to experience complementarity, for only in this case it may happen that, after answering to some query, the automaton undergoes an irreversible transition, making it impossible for the experimenter to probe a different observable (and *vice versa*). Reversibility does not change the picture, since if both the observer and the observed object were immersed in a reversible environment, then the experiment could be "undone" and the original automaton state reconstructed only at the price of loosing all the information gathered so far [40]. This is an analogue to the quantum eraser experiment [41] and other setups (e.g., [42]) developed for demonstrating the feasibility of a reconstruction of quantum states.

Bear in mind that complementarity is not only a feature of exotic finite models which were specially crafted for this particular purpose. Since these finite models represent a subset of objects that can be simulated by any universal computer, such as a CA or a Turing machine, complementarity is, in a sense, a generic and robust property of all computational universes.

2.2 Intrinsic undecidability

The quest to translate Gödel-Turing type recursion theoretic undecidability into physics has a long history. Gödel himself did not believe that his results have any relevance for physics, at least not for quantum physics[13]. Early on, Popper speculated about limits of forecast in the light of these findings [44]. More recent undecidability results are based on physical configurations which are provably unsolvable through the reduction to the halting problem (e.g., [45, 46, 47]).

Indeed, since the Turing machine is modelled after a paper and pencil real world scenario, universal computers can be embedded into certain physical systems capable of universal computation. Undecidabilities can then be obtained almost as a "free lunch;" i.e., by reduction to the recursive unsolvability

---

[13]In [43, 140-141], Bernstein writes, *Wheeler said, "I went to Gödel, and I asked him, 'Prof. Gödel, what connection do you see between your incompleteness theorem and Heisenberg's uncertainty principle?' I believe that Wheeler exaggerated a little bit now. He said, 'And Gödel got angry and threw me out of his office!' Wheeler blamed Einstein for this. He said that Einstein had brain-washed Gödel against quantum mechanics and against Heisenberg's uncertainty principle!* (The author has asked professor Wheeler and got this anecdote confirmed.)

of certain prediction problems, such as the halting problem.

So, why do people such as Casti[14], who had been very interested in the subject, consider this issue as a "red herring?" Maybe because so far not a single problem of relevance in physics not constructed for this particular purpose is provably undecidable.

2.3 Continuum *versus* discrete physics

The conceptualization of the number system—from just a few finger counts to the natural numbers, the integers, rationals, reals [50] and further on to complex numbers, quaternions and hyperreals is undoubtedly one of the most beautiful and greatest achievements of humanity. Nevertheless, as more and more abstractions enter these great patterns of thought, one is compelled to question their practical physical relevance. Clearly, for instance, infinite divisibility (from the rational onwards) and continuity (from the reals onwards) find strong pragmatic justifications by their applicability to almost all branches of theoretical physics, including quantum mechanics. Even so, some doubts as to the appropriateness of transfinite concepts in physical modelling remain [51]. Let us state the following correspondence principle between physical phenomena and their models [52]: *every feature of a computational model should be reflected by the capacity of the corresponding physical system. Conversely, every physical capacity, in particular of a physical theory, should correspond to a feature of an appropriate computational model.*

Nature does not seem to allow Zeno squeezing [53, 54, 55, 56, 57, 58, 59, 29, 33] and other transfinite processes. It could therefore be conjectured that, as physical systems do not possess adequate transfinite capacities, only finite computational models ought to be acceptable for theoretical modelling. This still admits universal computation and finite automata, but it wipes out classical, nonconstructive continua.

Having said this, there may be some indication of absolute randomness involved in certain quantum measurements, though. Suppose a single electron is prepared in a particular spin state in one direction θ_p. Assume further that its spin state is not measured in this particular direction, but in another direction θ_m. Then quantum mechanics predicts that the probability that identical spin states are measured is $\cos^2[(\theta_p - \theta_m)/2]$; for a non-identical result the probability is $1 - \cos^2[(\theta_p - \theta_m)/2] = \sin^2[(\theta_p - \theta_m)/2]$ (classically, one would expect linear dependencies on the measurement angles, such as $1 - |\theta_p - \theta_m|/\pi$

---

[14]Casti (co-)organized two conferences; one in Santa Fe [48] and one in Abisko [49], bringing together many who were interested in this issue at the time.

and $|\theta_p - \theta_m|/\pi$, respectively). Moreover, quantum mechanics postulates that these outcomes are stochastic and cannot be reduced to some form of microscopic law governing the single measurement outcomes. At $\theta_p - \theta_m = \pi/2$, a series of such experiments, when coded into a 0, 1-sequence, is postulated by the quantum mechanical canon to render an algorithmically incompressible random sequence [60]; a fact which can be used to construct a plug-in device [61]; just like another card which can be inserted into a computer and facilitates the desired function, in this case the production of random data.

Here seems to be a physical source of absolute randomness [24, 25, 26] which appears almost totally "free" of any computational costs. Just detune preparation and measurement to attain the goal of a perfect random number generator. Indeed, this quantum postulate of microphysical randomness seems to be a remarkable fact, in particular since randomness is a valuable resource which, in the context of universal computation, cannot be obtained "for nothing." Although "almost all" reals are random, it is hard (indeed impossible) to come by any concrete element with that property. The closest one could get may be Chaitin's Ω number, which is the Kraft sum of the length of all prefix-free halting programs on some universal computer (see [24, 25, 26] for details). It is even possible to write down a finite program for computing the first bits of Ω, but for better precision there is no computable radius of convergence certifying that a particular finite sequence is the starting sequence for Ω. In that respect Ω resembles Specker's sequence of rational numbers with non-recursive limit, or the Busy Beaver function [62, 63, 64].

So, is every electron a point particle capable of transfinite computations? While electrons do not seem to possess any capacity of universal computation at all, they appear just to be perfect random number generators. That is indeed amazing! Maybe we just have not listened carefully enough when crafting the computational models appropriate for physics. Is Turin's universal computer model, appended with an additional "random oracle" plug-in, sufficient?

In another scenario, an electron might just be coded to carry the answer to a single question; e.g., related to its spin state in a particular direction. If requested to answer a different question, such as about its spin state in a different direction, it might just churn out random nonsense [65] according to Malus' law [66, 67]. Or, it may need an interface, an environment or measurement apparatus translating the observer's question to the language (or question) understandable by the object [40, 34], thereby introducing stochastic noise by uncontrollable macroscopic processes. So far, these are all metaphysical speculations which need to be sorted out by operational means, i.e., by experiment.

2.4 Nonlocality & Contextuality

Quantum nonlocality is a phenomenon which can be quite easily described, yet remains mysterious. Consider again the spin state measurements of electrons. Let us assume that it is possible to produce two particles in a singlet state, such that, when their spin is measured along an arbitrary but identical direction, their spin states are opposite. Now, consider the correlation of their spin states when measured along arbitrary but different directions. As it turns out, if the directions are different from 0 and from integer multiples of $\pi/2$ or π, the quantum correlations are either weaker or stronger than the classical correlations. This is related to the difference of the aforementioned quantum probabilities *versus* the classical ones. In terms of elementary physical events, one obtains more or fewer joint clicks in the detectors measuring the spin states than would be conceivable classically for any such state. The doctrine of "peaceful coexistence" between relativity theory and quantum mechanics [68] assures that this feature cannot be used for faster than light quantum signalling [69, 70, 71].

Can CA with local neighbourhood cell evolution reproduce quantum-type nonlocality? That seems to be a hard problem, in particular if one clings to the idea of evolution functions which only depend on the neighbourhood, a property which surely seems to be a constituent element in the definition of CA. Indeed, with regards to nonlocality, little convincing evidence and comfort has been given so far by the CA community. Zuse mentions the chess metaphor of the bishop, a piece which can move in single-colour diagonal direction only, thereby exerting a nonlocal influence on the entire chessboard [16]. But how could the entire chessboard know of the bishop's motion if information can only propagate by one cell per time step? Considering quantized cells is no solution, because the quantum nonlocalities introduced by proper normalization of the entire ray wave function comes as no surprise [72]: quantum behaviour of a quantized system is indeed to be expected.

Another, entirely different and radical possibility would be to give up the notion of "calculating space" and consider a computational substratum which is nonlocal from the very beginning. In this approach, the cellular space does not correspond to anything which is spatially extended from a physical point of view, such as the tesselated configuration space Zuse had in mind. Rather, it might be some kind of generalized phase space, in which physical states are discrete. This resembles the "old" quantum mechanics of Planck and

Einstein[15] and known as Bohr-Sommerfeld "quantization."

Contextuality is another controversial issue which is discussed in the quantum context [74]. One may argue that as it cannot be operationalized anyway, contextuality is a property of almost pure theoretical value, such as counterfactuals or scholastic *infuturabilities*. In continuum theory, there are "exotic" ways to come by [75, 76], but this is no option for a discrete model. At first sight, classical computers seem to be value definite and noncontextual, but a closer inspection reveals that there are subtleties to be kept in mind. Value definiteness need not imply that an agent is prepared to answer *any* experimental question. Indeed, in contradistinction to Kant's transcendental ideal[16], and scholastic, theological speculations whether or not the omniscience of God extends to events which would have occurred if something had happened that did not happen, which have been so powerfully formalized into a finitistic proof (cf. [78, p. 243] and [79, p. 179]), some properties may not be properly definable for certain computational agents, and therefore may not be operational. For example, if an agent trained to wash dishes is confronted with the task to write a book on hiking trails in New Zealand's Waitakere ranges, it will most certainly be at a complete loss. Or an agent advised to direct some parties to a path on the right hand side when asked for right or left, will most certainly be at a loss when asked whether to proceed up or down. The agent simply would not be programmed and thus not be prepared to answer any type of question, but rather only a small selection from among all conceivable ones.

3 Intrinsic, Embedded Observer Mode

Computational complementarity and undecidability in general are good examples of how the science of systems may enter physics. Unless one accepts the concept of an "intrinsic embedded mode," computational complementarity disappears into thin air. And since system science seems foreign to most physicists, it is hard to see if and when such concepts will be more broadly recognized.

---

[15]As expressed by Planck [73, p. 387], *"Again it is confirmed that the quantum hypothesis is not based on energy elements but on action elements, according to the fact that the volume of phase space has the dimension h^f."*

[16]In the *3. Hauptstück, 3. Abschnitt. Von dem transzendentalen Ideal (Prototypon transscendentale)*, of *"Kritik der reinen Vernunft,"* [77], Kant states, "But again, everything, as regards its possibility, is also subject to the principle of complete determination, according to which one of all the possible contradictory predicates of things must belong to it." The German original reads, *"Ein jedes Ding aber, seiner Möglichkeit nach, steht noch unter dem Grundsatze der durchgängigen Bestimmung, nach welchem ihm von allen möglichen Prädikaten der Dinge, sofern sie mit ihren Gegenteilen verglichen werden, eines zukommen muß."*

As with all general concepts, it is hard to pinpoint when exactly the concept of an intrinsic embedded observer was formulated for the first time[17]. Boskovich [80] around 1755 referred to the fact that embedded observers cannot recognize an overall change (squeeze, dilatation and contraction) of the system size[18]. More recently, Toffoli [81] discussed the role of the observer in uniform systems. Embedded observers are *the* big issue in relativity theory, because Einstein insisted on operational methods available within the system only in defining clocks, length scales, and when comparing them[19]. Rössler [86, 87] and the author [82, 83, 84, 88, 29], independently share similar concepts, although Rössler's emphasis has been on the role of the interface between observer and observed object [89] rather than on concrete examples of automaton logic or space-time frames [20].

4 Space Time Frames of Intrinsic Observers

Relativity theory has altered the way we think of space and time from a formal point of view, but the perception of space and time at large, and what meaning is ascribed to these notions, has not changed too much: while pre-relativistic "Galilean" type thinking considered space and time as absolute and immutable, nowadays this role is ascribed to the relativistic forms of space time coordinates and their transformation laws. It is almost as if the attitude of the protagonists remained the same, but their message changed slightly.

---

[17]One is also tempted to mention Archimedes' *"points outside the world from which one could move the earth."* Mind that Archimedes' use of "points outside the world" was in a mechanical rather than in a metatheoretical context: he claimed to be able to move any given weight by any given force, however small.

[18]In [80], Boskovich states, *"... And we would have the same impressions if, under conservation of distances, all directions would be rotated by the same angle, ... And even if the distances themselves would be decreased, whereby the angles and the proportions would be conserved, ...: even then we [[the observers]] would have no changes in our impressions. ... A movement, which is common to us [[the observers]] and to the Universe, cannot be observed by us; not even if everything would be stretched or shrinked by an arbitrary amount."*

[19]The author had some problem to publish a paper on embedded observers in relativity theory, apparently because of the rather unconventional nature of the subject. However, after an appeal, the paper became preprint #LBL-16097 [82] and was later published in a revised version [83] (see also [84]). I write this to encourage young researchers not to give up in pursuing their own nonfashionable ideas [85].

[20]When I was invited to participate to a Linz *Ars Electronica* conference on "Endophysics" in 1992, I was almost shocked by the similarity between Rössler's thoughts and the ones I had pursued. One follow-up meeting was organized by Atmanspacher in Germany [90]. I later learned that for researchers trained in mathematical system science, like John Casti, embedded observers sounded like a very familiar, almost self-evident concept. [91, 92, 93].

Relativity theory, as introduced by Einstein, at least in the first, kinematic, part of the seminal 1905 paper [94], is conceived as a strictly operational theory for embedded, intrinsic observers. Those observers are bound to use the methods and capabilities of the system of which they are an integral part; and they cannot resort to an extrinsic, "God's eye" overview of it. But operationalism is not enough to create space-time frames. What is also needed (but seldom mentioned although implicitly assumed) are conventions for measuring time and space, and for comparing those scales at different locations and different times in co-moving and other experimental configurations. Indeed, the International System of units outrightly declares a previously experimental physical fact to be convention. The speed of light is assumed to be constant for all reference frames. With the mild side assumption of the one-to-oneness (invertibility) of space-time transformation, this convention declares the preservation of light cones, and thus, by the preservation of set theoretic intersections of light cones such as time- space- and lightlike onedimensional subspaces, results in affinity and linearity of the transformation laws. From this point of view, the Lorentz transformation is a geometric statement, not a physical entity. In geometry, this is known as Alexandrov's theorem [95, 96, 97, 98, 99, 100, 101].

So, in a sense, this is the big picture. If one requires invariance of some "fundamental" speed and bijectivity of the transformation laws, then the Lorentz-type transformation laws containing that "fundamental" speed follow. Thereby, it makes no difference whether the associated observers are embedded in the "real" Universe, in a CA, or in a plum pudding; as long as these conditions and conventions are met, then Alexandrov's theorem certifies that the geometry is a relativistic one. For discrete models, these results will always be only approximations which are valid down to scales where the discreteness becomes important.

Where is all the physics gone? The answer to this question is that the physics is in the invariance with respect to any such Lorentz-type transformations. For example, clocks governed by electromagnetic phenomena will be showing the "right" time in all frames if the "fundamental" speed is chosen to be the speed of light. Sound clocks tick invariantly in the respective system if the "fundamental" speed is the speed of sound. Scales are invariant if the forces stabilizing that scales are electromagnetic ones and the "fundamental" speed is again chosen to be the speed of light. So, with these new conventions, the invariance of certain length [102] and time scales, corresponding to the relativistic form invariance of the laws governing them, becomes a physical statement [103, 102, 104, 105].

The following is an example [106] of an Einstein synchronisation by clocks generating radar coordinates in a one-dimensional CA with the following evolution rules.

$$\begin{array}{lllll}
\varphi(>,\_,X) \to >, & \varphi(X,\_,<) \to <, & \varphi(\_,\_,\_) \to \_, & \varphi(X,\_,>) \to \_, & \varphi(<,\_,X) \to \_, \\
\varphi(\_,>,\_) \to \_, & \varphi(\_,<,\_) \to \_, & \varphi(\_,>,I) \to <, & \varphi(I,<,\_) \to >, & \varphi(>,I,X) \to *, \\
\varphi(<,*,X) \to I, & \varphi(X,<,*) \to \_, & \varphi(*,1,X) \to 2, & \varphi(*,2,X) \to 3, & \varphi(*,3,X) \to 4, \\
\varphi(*,4,X) \to 5, & \varphi(*,5,X) \to 6, & \varphi(*,6,X) \to 7, & \varphi(*,7,X) \to 8, & \varphi(*,8,X) \to 9, \\
\varphi(*,9,X) \to 0, & \varphi(*,0,X) \to 1, & \varphi(0,\_,X) \to \_, & \varphi(1,\_,X) \to \_, & \varphi(2,\_,X) \to \_, \\
\varphi(3,\_,X) \to \_, & \varphi(4,\_,X) \to \_, & \varphi(5,\_,X) \to \_, & \varphi(6,\_,X) \to \_, & \varphi(7,\_,X) \to \_, \\
\varphi(8,\_,X) \to \_, & \varphi(9,\_,X) \to \_, & \varphi(X,*,0) \to *, & \varphi(X,*,1) \to *, & \varphi(X,*,2) \to *, \\
\varphi(X,*,3) \to *, & \varphi(X,*,4) \to *, & \varphi(X,*,5) \to *, & \varphi(X,*,6) \to *, & \varphi(X,*,7) \to *, \\
\varphi(X,*,8) \to *, & \varphi(X,*,9) \to *, & \varphi(X,1,X) \to 1, & \varphi(X,2,X) \to 2, & \varphi(X,3,X) \to 3, \\
\varphi(X,4,X) \to 4, & \varphi(X,5,X) \to 5, & \varphi(X,6,X) \to 6, & \varphi(X,7,X) \to 7, & \varphi(X,8,X) \to 8, \\
\varphi(X,9,X) \to 9, & \varphi(X,0,X) \to 0, & \varphi(X,I,0) \to I, & \varphi(X,I,1) \to I, & \varphi(X,I,2) \to I, \\
\varphi(X,I,3) \to I, & \varphi(X,I,4) \to I, & \varphi(X,I,5) \to I, & \varphi(X,I,6) \to I, & \varphi(X,I,7) \to I, \\
\varphi(X,I,8) \to I, & \varphi(X,I,9) \to I, & \varphi(X,I,0) \to I, & \varphi(X,I,X) \to I, & \varphi(*,\_,X) \to \_, \\
\varphi(\_,\_,I) \to \_, & \varphi(I,\_,\_) \to \_ & \varphi(I,>,\_) \to \_, & \varphi(\_,<,I) \to \_. &
\end{array}$$

Here, X stands for any state except the ones already specified. These rules look a little bit "murky," but they can be simulated by any universal CA and they serve their purpose to demonstrate clock synchronization procedures.

Assume two clocks at two arbitrary points A and B in the CA which are "of similar kind." At some arbitrary A-time t_A a ray goes from A to B. At B it is instantly (without delay) reflected at B-time t_B and reaches A again at A-time $t_{A'}$. The clocks in A and B are *synchronized* if $t_B - t_A = t_{A'} - t_B$. The two-ways ray velocity is given by $2|AB|/(t_{A'} - t_A) = c$, where $|AB|$ is the distance between A and B. In Fig. 1(a), an example of synchronization between two clocks A and B is drawn.

What happens with the intrinsic synchronization and the space-time coordinates when observers are considered which are in motion with respect to the CA? For simplicity, suppose a constant motion of v automaton cells per time cycle. With these units, the ray speed is $c = 1$, and $v \leq 1$. There are numerous ways to simulate sub-ray motion on a CA. In what follows, the case $v = 1/3$ will be studied in such a way that every three CA time cycles the walls, symbolised by I, move one cell to the right.

Notice that two clocks which are synchronized in a reference frame which is at rest with respect to the CA medium are *not synchronized* in their own co-moving reference frame. Consider, as an example, the CA drawn in Fig. 1(b). (Strictly speaking, the CA rule here depends on a two-neighbor interaction.) For $t_A = 1$, $t_B = 4$, $t_{A'} = 5$, and $4 - 1 \neq 5 - 4$, if the first clock is corrected to make up for the different time of ray flights as in Fig. 1(c), $t_A = 2$, $t_B = 4$, $t_{A'} = 6$, and $4 - 2 = 6 - 4$. Then, this correction amounts to an asynchronicity of the two ray clocks with respect to the "original" CA medium.

```
   clock1   A        B   clock2          clock1   A        B   clock2          clock1   A        B   clock2
_____I>__I0__I>_____I__I>__I0_____    _I>_I0__I___<__I__I>_I0_____  _I>_I1__I___<__I__I>_I0_____
____I_>_I0__I_>_____I__I_>_I0_____    _I_>I0__I__<____I__I_>I0_____  _I_>I1__I__<____I__I_>I0_____
_____I__>I0__I_>____I__I__>I0_____    _I_<*0__I_<_____I__I_<*0_____  _I_<*1__I_<_____I__I_<*0_____
_____I__<*0__I_>____I__I__<*0_____    __I>_I1__I_>_____I__I>_I1_____  _I>_I2__I>_____I__I>_I1_____
_____I_<_I1__I_____>_I__I_<_I1_____    __I_>I1__I_>____I__I_>I1_____  _I_>I2__I_>_____I__I_>I1_____
_____I<__I1__I_____>_I__I<__I1_____    __I_<*1__I_>____I__I_<*1_____  __I_<*2__I_>____I__I_<*1_____
____I>__I1__I_>_____I__I_>_I1_____    ___I>_I2__I_>___>__I__I_>I2_____  __I>_I3__I_>____>__I__I_>I2_____
____I_>_I1__I_____<*__I_>_I1_____    ___I_>I2__I_>____>__I__I_>I2_____  __I_>I3__I_>____>__I__I_>I2_____
_____I__>I1__I_____<_I__I__>I1_____    ___I_<*2__I___>___>__I__I_<*2_____  __I_<*3__I____>__I__I_<*2_____
_____I__<*1__I____<___I__I_<*1_____    ____I>_I3__I___>__I__I_>I3_____  ____I>_I4__I___>__I__I_>I3_____
____I_<_I2__I___<___I__I_<_I2_____    ____I_>I3__I____>_I__I_>I3_____  ____I_>I4__I____>_I__I_>I3_____
____I<__I2__I_<_____I__I_<__I2_____    ____I_<*3__I_____>I__I_<*3_____  ____I_<*4__I_____>I__I_<*3_____
____I>__I2__I_<_____I__I>_I2_____    _____I>_I4__I_____>I__I>_I4_____  _____I>_I5__I_____>I__I>_I4_____
____I_>_I2__I<_____I__I_>I2_____    _____I_>I4__I_____<*__I_>I4_____  _____I_>I5__I_____<*__I_>I4_____
____I__>I2__I_____I__I__>I2_____    _____I_<*4__I_____<_I__I_<*4_____  _____I_<*5__I_____<_I__I_<*4_____
____I__<*2__I_>_____I__I__<*2_____    _____I>_I5__I___<___I__I_>I5_____  _____I>_I6__I___<___I__I_>I5_____
____I_<_I3__I_>_____I__I_<_I3_____    _____I_>I5__I___<____I__I_>I5_____  _____I_>I6__I___<____I__I_>I5_____
____I<__I3__I_____I__I_<__I3_____    _____I_<*5__I_<_____I__I_<*5_____  _____I_<*6__I_<_____I__I_<*5_____
____I>__I3__I_____>_I__I>_I3_____    _____I>_I6__I>_____I__I_>I6_____  _____I>_I7__I>_____I__I_>I6_____
____I_>_I3__I_____>_I__I_>I3_____    _____I_>I6__I_>_____I__I_>I6_____  _____I_>I7__I_>_____I__I_>I6_____
____I__>I3__I_____>_I__I__>I3_____    _____I_<*6__I_>____I__I_<*6_____  _____I_<*7__I_>____I__I_<*6_____
____I__<*3__I____<*__I__I_<*3_____    _____I>_I7__I_>___I__I_>I7_____  _____I>_I8__I_>___I__I_>I7_____
____I_<_I4__I___<__I__I_<_I4_____    _____I_>I7__I_>__I__I_>I7_____  _____I_>I8__I_>___I__I_>I7_____
____I<__I4__I__<__I__I_<__I4_____    _____I_<*7__I__>__I__I_<*7_____  _____I_<*8__I__>__I__I_<*7_____
____I>__I4__I_<_____I__I>_I4_____    _____I>_I8__I_____>I__I>_I8_____  _____I>_I9__I_____>I__I>_I8_____
____I_>_I4__I_<_____I__I_>I4_____    _____I>_I8__I_>__I__I_>I8_____  _____I_>I9__I_>__I__I_>I8_____
____I__>I4__I<_____I__I__>I4_____    _____I_<*8__I___>I__I_<*8_____  _____I_<*9__I___>I__I_<*8_____
____I__<*4__I<_____I__I__<*4_____    _____I>_I9__I____>I__I_>I9____  _____I>_I0__I____>I__I_>I9_____
____I_<_I5__I>_____I__I_<_I5_____    _____I>_I9__I____<*__I_>I9____  _____I_>I0__I____<*__I_>I9_____
____I<__I5__I_>_____I__I_<__I5_____    _____I_<*9__I___<_I__I_<*9____  _____I_<*0__I___<_I__I_<*9_____
____I>__I5__I_>_____I__I>_I5_____    _____I>_I0__I___<___I__I_>I0__  _____I>_I1__I___<___I__I_>I0____
____I_>_I5__I_>_____I__I_>I5_____    _____I_>I0__I__<____I__I_>I0_  _____I_>I1__I__<____I__I_>I0____
____I__>I5__I_>____I__I__>I5_____    _____I_<*0__I_<_____I__I_<*0_  _____I_<*1__I_<_____I__I_<*0____
____I__<*5__I____>__I__I_<*5_____    _____I>_I1__I_>_____I__I_>I1  _____I>_I2__I_>_____I__I_>I1___
____I_<_I6__I___>__I__I_<_I6_____    _____I_>I1__I_>____I__I_>I1_  _____I_>I2__I_>____I__I_>I1____
____I<__I6__I_____>I__I_<__I6_____    _____I_<*1__I_>____I__I_<*1_  _____I_<*2__I_>____I__I_<*1____
____I>__I6__I___<__I__I_>I6_____    _____D_I2__I_>___>__I__I_>I2 _____D_I3__I_>___>__I__I_>I2___
____I_>_I6__I_<____I__I_>_I6_____    _____I>_I2__I_>___>__I__I_>I2 _____I>_I3__I_>___>__I__I_>I2__
____I__>I6__I_<____I__I__>I6_____    _____I_<*2__I____>__I__I_<*2 _____I_<*3__I____>__I__I_<*2___
____I__<*6__I_<____I__I__<*6_____    _____I>_I3__I___>__I__I_>I3 _____I>_I4__I___>__I__I_>I3___
____I_<_I7__I<_____I__I_<_I7_____    _____I_>I3__I____>_I__I_>I3 _____I_>I4__I____>_I__I_>I3___
```

(a)   (b)   (c)

Figure 1: Synchronization by ray exchange (a) in a system as rest with respect to a CA; (b) ray exchange with synchronization defined by (a); (c) synchronization in co-moving frame.

5 Now What?

Despite all these efforts, including those of the author presented above, the computational approach to understanding universes has so far resulted in little or no phenomenological impact; not to speak of any "killer application" which would make the few critics and the many hesitant researchers listen to the subject. Large segments of theoretical physics appear to be in the very same position in other areas such as string theory or quantum gravity as well, but this is no big comfort. In search for applications of the idea of computational universes let us shortly discuss some of the predictions of the subject and their possible empirical validation or falsification.

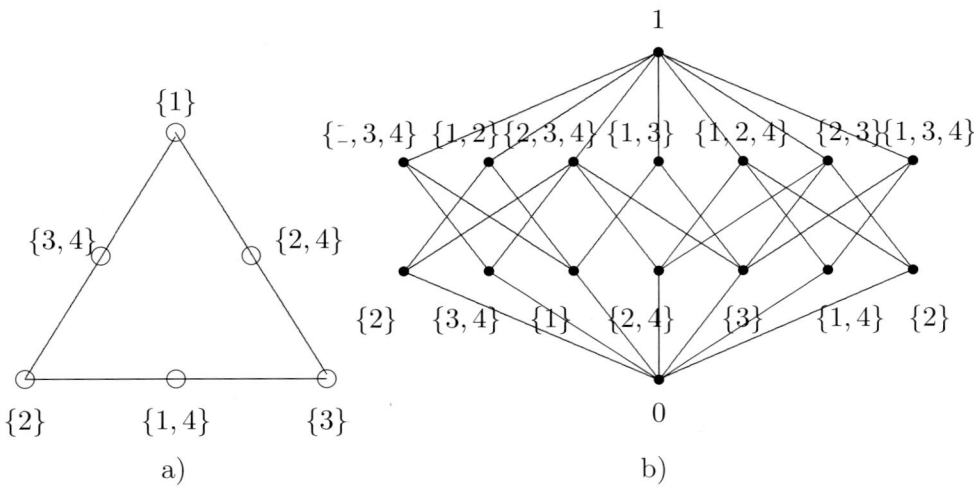

Figure 2: a) Greechie and b) Hasse diagram of a logic featuring complementarity which is not a quantum logic but which is embeddable in a Boolean logic.

5.1 New range of phenomena

With regards to the logical order of propositions, there may exist phenomena perceivable by intrinsic, embedded observers which cannot happen according to quantum mechanics but are realizable by finite automata. The simplest case is characterized by a Greechie hyperdiagram of triangle form, with three atoms per edge. Its automaton partition logic is given by

$$\{\{\{1\},\{2\},\{3,4\}\},\{\{1\},\{2,4\},\{3\}\},\{\{1,4\},\{2\},\{3\}\}\}. \qquad (1)$$

A corresponding Mealy automaton is $\langle\{1,2,3,4\},\{1,2,3\},\{1,2,3\},\delta=1,\lambda\rangle$, where $\lambda(1,1) = \lambda(3,2) = \lambda(2,3) = 1$, $\lambda(3,1) = \lambda(2,2) = \lambda(1,3) = 2$, and $\lambda(2,1) = \lambda(4,1) = \lambda(1,2) = \lambda(4,2) = \lambda(3,3) = \lambda(4,3) = 3$.

Figure 2 depicts the Greechie and Hasse diagrams of this propositional structure. The physical interpretation of Eq. (1) is the following: there exist six observables $\{1\}$, $\{2\}$, $\{3\}$, $\{1,4\}$, $\{2,4\}$, and $\{3,4\}$; i.e., $\{3,4\}$ corresponds to " *the system is in state 3 or in state 4.* "

They are grouped into three partitions of $\{1,2,3,4\}$, such that within each group the observables are comeasurable. For instance, in the automaton example enumerated above, the experiment with the input of symbol 2 differentiates between $\{1,4\}$, $\{2\}$, $\{3\}$ and all properties obtained by forming the logical "or" operation, such as $\{2,3\}$. But the experiment does not reveal all conceivable propositions, such as $\{1,2\}$. Another experiment with the input of symbol 3 does, but cannot reveal other properties, such as $\{1,4\}$. Because of this complementarity, the propositions are nonclassical, in particular they do not obey the distributive law: since $\{1,3\} \vee \{2,3\} = \{1,2,3,4\}$,

$$\begin{aligned}\{1,2,4\} \wedge (\{1,3\} \vee \{2,3\}) = \{1,2,4\} \wedge \{1,2,3,4\} &= \{1,2,4\} \\ = (\{1,2,4\} \wedge \{1,3\}) \vee (\{1,2,4\} \wedge \{2,3\}) &= \{1,2\}.\end{aligned}$$

This humble propositional structure is thus non-classical, but quite remarkably it also cannot be realized by quantum mechanics. The complementary groups are interlocked in a triangle form, which is forbidden by the Hilbert space based algebraic structure of quantum mechanics: In analogy to Kochen and Specker [107], we denote by the symbol "$\perp$" the binary relation of comeasurability. Any sequencing of observables such as

$$\{1\} \perp \{3,4\} \perp \{2\} \perp \{1,4\} \perp \{3\} \perp \{2,4\} \perp \{1\}$$

(with $\{1\} \not\perp \{1,4\} \not\perp \{2,4\}$ and so on) cannot occur in quantum mechanics. Hence, if this propositional structure is experienced in some physical setup, then quantum mechanics is not an appropriate theoretical representation for it. Computational universes would be a natural candidate.

5.2 Coarse grained structure of digital space

Already Zuse mentioned that, if space is tesselated, then this tesselation will eventually show up; either by some anisotropy or by a fundamental length scale. No indication is given exactly when this granularity should show up; and problems abound [108]. Yet, there is no guarantee that space and time will be organized as a regular lattice; it may rather resemble a huge pile of more or less randomly and densely packed sand and stabilized by whatever forces there are.

In view of the mild discreteness of quantum mechanics already mentioned earlier (only an integer number of quanta per field node), it might well be that we have already unravelled the fundamental discreteness; but not in the properties where we had expected them. So, maybe the field nodes or phase space are more fundamental than the frames of space and time that we use to define those fields. In this idealistic picture, space and time may be convenient constructions of our minds to sort out the evolution of field modes.

5.3 Exotic probabilities

One approach to the formalism is that anything which is not forbidden explicitly is realized[21]. As Gleason's theorem strongly ties quantum probabilities to Hilbert space, there may be non-classical and non-quantum probabilities which can be modelled with automaton or generalized urn models.

Let us consider again spin state measurements on electrons modelled by two-dimensional Hilbert space entities. The associated algebra of propositions consists of (the horizontal sum of) Boolean sublattices 2^2 which are pasted together [109] at their extreme elements. In this case, Gleason's theorem does not apply. By taking the algebraic structure and the set of dispersion free (two-valued) states alone, there exists the possibility of nongleason type probability measures. These measures have singular, separating distributions and thus can be embedded into "classical" Boolean algebras such as generalized urn und automaton partition logics. One particular example is represented in Figure 3. Its probability measure is $P(x_-^i) = 1$ and $P(x_+^i) = 1 - P(x_-^i) = 0$

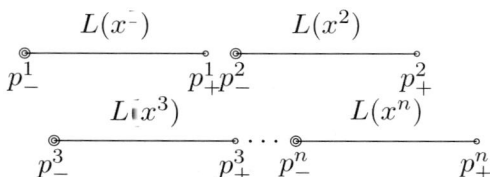

Figure 3: Example for a nongleason type probability measure for n spin one-half state propositional systems $L(x^i), i = 1, \cdots, n$ which are not comeasurable. The superscript i represents the ith measurement direction. The concentric circles indicate the atoms with probability measure 1.

for $i = 1, \ldots, n$. The associated automaton models are straightforward. Every such dispersion free state is obtained by associating with it a particular automaton state. Whether or not such probability distributions exist for fundamental processes is an open question. For spin state measurements of the electrons, this does not seem to be the case, but again the question of state preparation may be essential here.

Another more exotic example of a suborthoposet which is embeddable into the three-dimensional real Hilbert lattice $C(\mathbb{R}^3)$ and can also be realized by generalized urn models and finite automata has been presented by Wright [36]. Its Greechie diagram of the pentagonal form is drawn in Figure 4. Wright

---

[21]Feynman's rule of thumb states that whatever is not explicitly forbidden is mandatory. See also the "go-go" principle introduced in [52].

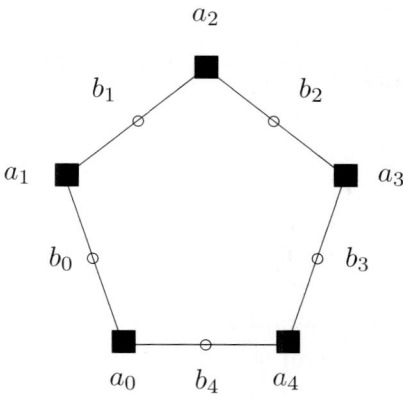

Figure 4: Greechie diagram of the Wright pentagon [36]. Filled squares indicate probability $\frac{1}{2}$.

showed that the probability measure $P(a_i) = \frac{1}{2}$, $P(b_i) = 0$, for $i = 0, 1, 2, 3, 4$, as depicted in Figure 4, is no convex combination of other pure states; and furthermore, that it does not correspond to any Gleason type measure allowed as quantum probability. In this sense, it is a "stranger-than-quantum probability." And although automata and generalized urn models ass well as quantum system with this pentagonally interlocking algebraic structure of propositions exist, no realizable probability measure on it is of the form of Wright's measure defined above. The reason for this is the impossibility to represent it as a convex combination of other dispersion free two-valued states.

5.4 "Tuning" reality

If the physical phenomena are the intrinsic view of a mathematical or computational universe, then any attempt to render, manipulate and change certain phenomena could be interpreted as "reprogramming." In fact, reprogramming or "tuning"[22] reality may be a powerful new metaphor hitherto foreign to theoretical physics. Again, one should keep in mind that this is highly speculative.

5.5 Against odds

Let me again emphasise that discrete or algorithmic physics may be utterly non mainstream and off-topic, as competing with "traditional" continuum physics

---

[22]The term "tuning" is borrowed from the movie *Dark City* by Alex Proyas, where similar motives have been casted.

is hard. For instance, note the fabulous coincidence between the theoretical and the experimental values of the anomalous magnetic moment of the Muon $a_{\mu,t} = 11659177(7) \times 10^{-10}$ and $a_{\mu,e} = 11659204(7)(5) \times 10^{-10}$ [110]. Or take the neutron double slit experiments [111] which show a wonderful agreement of theory and experiment.

Yet, despite all these difficulties, discrete computational physics certainly represents an interesting, speculative and challenging research area. Many ideas from system science, interface design, to dualism (e.g., the Eccles Telegraph) enter. The issue has metaphysical connotations. It is for instance not totally unreasonable that a demiurge would create an "atomistic" world such as ours, in which an immense (to us) number of discrete gaming pieces come together to form a universe and which are constantly rearranged to form rich and varied and seemingly complex patterns. Or there is just one consistent Universe of Mathematics, and this is the physical Universe we are living in.

Acknowledgements

Many thanks go to Tim Boykett for inviting me to a Time's Up workshop, giving me the feeling that somewhere out there are people still interested and listening. Garry J. Tee from the University of Auckland has provided reference [4], and Peter Mittelstaedt referred to the beautiful quotation of Kant. I am particularly thankful to Ross Rhodes for his continuing encouragement; also for reading an earlier version of the manuscript and for many suggestions to improve the text.

References

[1] J. C. Eccles, The mind-brain problem revisited: The microsite hypothesis, in: J. C. Eccles, O. Creutzfeldt (Eds.), The Principles of Design and Operation of the Brain, Springer, Berlin, 1990, p. 549.

[2] R. Descartes, Meditation on First Philosophy, 1641,
URL http://oregonstate.edu/instruct/phl302/texts/descartes/%meditations/meditations.html

[3] H. Putnam, Reason, Truth and History, Cambridge University Press, Cambridge, 1981.

[4] J. M. Thomas, Michael Faraday and the Royal Institution, Adam Hilger, Bristol, 1991.

[5] T. Bulfinch (Ed.), The Age of Fable, 1913, project Gutenberg Etext #3327 & Etext #4928.

[6] C. B. Boyer, A History of Mathematics, John Wiley & Sons, New York, 1968.

[7] K. Gödel, in: S. Feferman, J. W. Dawson, S. C. Kleene, G. H. Moore, R. M. Solovay, J. van Heijenoort (Eds.), Collected Works. Publications 1929-1936. Volume I, Oxford University Press, Oxford, 1986.

[8] E. Fredkin, T. Toffoli, Conservative logic, International Journal of Theoretical Physics 21 (1982) 219–253.

[9] J. von Neumann, Theory of Self-Reproducing Automata, University of Illinois Press, Urbana, 1966, a. W. Burks, editor.

[10] E. P. Wigner, The unreasonable effectiveness of mathematics in the natural sciences. Richard Courant Lecture delivered at New York University, May 11, 1959, Communications on Pure and Applied Mathematics 13 (1960) 1.

[11] E. F. Moore, Gedanken-experiments on sequential machines, in: C. E. Shannon, J. McCarthy (Eds.), Automata Studies, Princeton University Press, Princeton, 1956.

[12] J. H. Conway, Regular Algebra and Finite Machines, Chapman and Hall Ltd., London, 1971.

[13] S. Ulam, Random processes and transformations, Proceedings of the International Congress of Mathematics 2 (1952) 264–275, talk held in 1950.

[14] K. Zuse, Rechnender Raum, Elektronische Datenverarbeitung (1967) 336–344. A scan is available at
URL http://www.idsia.ch/~juergen/digitalphysics.html

[15] K. Zuse, Rechnender Raum, Friedrich Vieweg & Sohn, Braunschweig, 1969, English translation as [112].

[16] K. Zuse, Discrete mathematics and Rechnender Raum, URL http://www.zib.de/PaperWeb/abstracts/TR-94-10/ (1994).

[17] A. Einstein, Grundzüge der Relativitätstheorie, 1st Edition, Vieweg, Braunschweig, 1956.

[18] C. H. Bennett, Logical reversibility of computation, IBM Journal of Research and Development 17 (1973) 525–532, reprinted in [113, pp. 197-204].

[19] T. Toffoli, N. Margolus, Invertible cellular automata: A review, Physica D 45 (1990) 229–253.
URL http://pm1.bu.edu/~tt/publ/ica.ps

[20] E. Fredkin, An informational process based on reversible universal cellular automata, Physica D45 (1990) 254–270, doi:10.1016/0167-2789(90)90186-S available at
URL http://www.digitalphilosophy.org

[21] S. Wolfram, A New Kind of Science, Wolfram Media, Inc., Champaign, IL, 2002.

[22] L. Gray, A mathematician looks at Wolfram's new kind of science, Notices of the American Mathematical Society 50 (2), URL http://www.ams.org/notices/200302/fea-gray.pdf.

[23] K. Gödel, A remark about the relationship between relativity theory and idealistic philosophy, in: P. A. Schilpp (Ed.), Albert Einstein, Philosopher-Scientist, Tudor Publishing Company, New York, 1949, pp. 555–561, reprinted in [114, pp. 202-207].

[24] G. J. Chaitin, Algorithmic Information Theory, Cambridge University Press, Cambridge, 1987.

[25] G. J. Chaitin, Exploring Randomness, Springer Verlag, London, 2001.

[26] C. Calude, Information and Randomness—An Algorithmic Perspective, 2nd Edition, Springer, Berlin, 2002.

[27] W. Brauer, Automatentheorie, Teubner, Stuttgart, 1984.

[28] D. Finkelstein, S. R. Finkelstein, Computational complementarity, International Journal of Theoretical Physics 22 (1983) 753–779.

[29] K. Svozil, Randomness & Undecidability in Physics, World Scientific, Singapore, 1993.

[30] M. Schaller, K. Svozil, Automaton logic, International Journal of Theoretical Physics 35 (1996) 911–940.

[31] C. Calude, E. Calude, K. Svozil, S. Yu, Physical versus computational complementarity I, International Journal of Theoretical Physics 36 (7) (1997) 1495–1523.

[32] K. Svozil, Quantum Logic, Springer, Singapore, 1998.

[33] K. Svozil, The Church-Turing thesis as a guiding principle for physics, in: C. S. Calude, J. Casti, M. J. Dinneen (Eds.), Unconventional Models of Computation, Springer, Singapore, 1998, pp. 371–385.

[34] K. Svozil, Finite automata models of quantized systems: conceptual status and outlook. In: M. Ito and M. Toyama (Eds.), Developments in Language Theory. Proceedings of DTL 2002. Springer, Berlin, 2003, pp. 93–102.
URL http://www.arxiv.org/abs/quant-ph/0209089

[35] K. Svozil, Logical equivalence between generalized urn models and finite automata (2002).
URL http://arxiv.org/abs/quant-ph/0209136

[36] R. Wright, The state of the pentagon. A nonclassical example, in: A. R. Marlow (Ed.), Mathematical Foundations of Quantum Theory, Academic Press, New York, 1978, pp. 255–274.

[37] R. Wright, Generalized urn models, Foundations of Physics 20 (1990) 881–903.

[38] L. Fortnow, One complexity theorists view of quantum computing, Theoretical Computer Science 292 (2003) 597–610.

[39] E. Bernstein, U. Vazirani, Quantum complexity theory, in: Proceedings of the 25th Annual ACM Symposium on Theory of Computing, San Diego, California, May 16-18, 1993, ACM Press, 1993, pp. 11–20.

[40] K. Svozil, Quantum interfaces, in: H. H. Diebner, T. Druckrey, P. Weibel (Eds.), Sciences of the Interface, Genista Verlag, Tübingen, Germany, 2001, pp. 76–88.
URL http://arxiv.org/abs/quant-ph/0001064

[41] T. J. Herzog, P. G. Kwiat, H. Weinfurter, A. Zeilinger, Complementarity and the quantum eraser, Physical Review Letters 75 (1995) 3034–3037.

[42] D. B. Greenberger, A. YaSin, "Haunted" measurements in quantum theory, Foundation of Physics 19 (1989) 679–704.

[43] J. Bernstein, Quantum Profiles, Princeton University Press, Princeton, 1991.

[44] K. R. Popper, Indeterminisim in quantum physics and in classical physics I,II. The British Journal for the Philosophy of Science 1 (1950) 117–133,173–195.

[45] A. Komar, Undecidability of macroscopically distinguishable states in quantum field theory, Physical Review 133 (1964) B542, URL: http://link.aps.org/abstract/PR/v133/pB542.

[46] I. Kanter, Undecidability principle and the uncertainty principle even for classical systems, Physical Review Letters 64 (1990) 332–335, URL: http://link.aps.org/abstract/PRL/v64/p332.

[47] C. D. Moore, Unpredictability and undecidability in dynamical systems, Physical Review Letters 64 (1990) 2354–2357, cf. Ch. Bennett, *Nature*, **346**, 606 (1990).
URL http://link.aps.org/abstract/PRL/v64/p2354

[48] J. L. Casti, Why undecidability is too important to be left to Gödel, in: J. L. Casti, J. F. Traub (Eds.), On Limits, Santa Fe Institute Report 94-10-056, Santa Fe, NM, 1994,
URL http://www.santafe.edu/sfi/publications/Working-Papers/94-10-056.pdf and http://www.santafe.edu/sfi/publications/wpabstract/199410056.

[49] J. L. Casti, A. Karlquist, Boundaries and Barriers. On the Limits to Scientific Knowledge, Addison-Wesley, Reading, MA, 1996.

[50] S. Drobot, Real Numbers, Prentice-Hall, Englewood Cliffs, New Jersey, 1964.

[51] P. W. Bridgman, A physicist's second reaction to Mengenlehre, Scripta Mathematica 2 (1934) 101–117, 224–234, cf. R. Landauer [115].

[52] K. Svozil, Set theory and physics, Foundations of Physics 25 (1995) 1541–1560.

[53] H. Weyl, Philosophy of Mathematics and Natural Science, Princeton University Press, Princeton, 1949.

[54] A. Grünbaum, Philosophical problems of space and time (Boston Studies in the Philosophy of Science, vol. 12), second, enlarged edition Edition, D. Reidel, Dordrecht/Boston, 1974.

[55] J. F. Thomson, Tasks and supertasks, Analysis 15 (1954) 1–13.

[56] P. Benacerraf, Tasks and supertasks, and the modern Eleatics, Journal of Philosophy LIX (24) (1962) 765–784.

[57] I. Pitowsky, The physical Church-Turing thesis and physical computational complexity, Iyyun 39 (1990) 81–99.

[58] M. Hogarth, Predicting the future in relativistic spacetimes, Studies in History and Philosophy of Science. Studies in History and Philosophy of Modern Physics 24 (1993) 721–739.

[59] J. Earman, J. D. Norton, Forever is a day: supertasks in Pitowsky and Malament-Hogart spacetimes, Philosophy of Science 60 (1993) 22–42.

[60] K. Svozil, The quantum coin toss—testing microphysical undecidability, Physics Letters A143 (1990) 433–437.

[61] T. Jennewein, U. Achleitner, G. Weihs, H. Weinfurter, A. Zeilinger, A fast and compact quantum random number generator, Review of Scientific Instruments 71 (2000) 1675–1680, arXiv:quant-ph/9912118.
URL http://xxx.lanl.gov/abs/quant-ph/9912118

[62] T. Rado, On non-computable functions, The Bell System Technical Journal XLI(41) (1962) 877–884, http://grail.cba.csuohio.edu/~somos/bb.html.

[63] A. H. Brady, The busy beaver game and the meaning of life, in: R. Herken (Ed.), The Universal Turing Machine. A Half-Century Survey, Kammerer und Unverzagt, Hamburg, 1988, p. 259.

[64] G. J. Chaitin, Computing the busy beaver function, in: T. M. Cover, B. Gopinath (Eds.), Open Problems in Communication and Computation, Springer, New York, 1987, p. 108, reprinted in [116].

[65] A. Zeilinger, A foundational principle for quantum mechanics, Foundations of Physics 29 (1999) 631–643.

[66] Č. Brukner, A. Zeilinger, Malus' law and quantum information, Acta Physica Slovaca 49 (1999) 647–652.

[67] Č. Brukner, A. Zeilinger, Information and fundamental elements of the structure of quantum theory, (2001).
URL http://www.arxiv.org/abs/quant-ph/0212084

[68] A. Shimony, Controllable and uncontrollable non-locality, in: S. K. *et al.* (Ed.), Proceedings of the International Symposium on the Foundations of Quantum Mechanics, Physical Society of Japan, Tokyo, 1984, pp. 225–230, see also J. Jarrett, *Bell's Theorem, Quantum Mechanics and Local Realism*, Ph. D. thesis, Univ. of Chicago, 1983; *Nous*, **18**, 569 (1984).

[69] N. Herbert, FLASH—a superluminal communicator based upon a new kind of quantum measurement, Foundation of Physics 12 (1982) 1171–1179.

[70] W. K. Wooters, W. H. Zurek, A single quantum cannot be cloned, Nature 299 (1982) 802–803.

[71] R. J. Glauber, Amplifiers, attenuators and the quantum theory of measurement, in: E. R. Pikes, S. Sarkar (Eds.), Frontiers in Quantum Optics, Adam Hilger, Bristol, 1986.

[72] S. Fussy, G. Grössing, H. Schwabl, Nonlocal computation in quantum cellular automata, Physical Review A 48 (1993) 3470–3477.
URL http://link.aps.org/abstract/PRA/v48/p3470

[73] M. Planck, Die physikalische Struktur des Phasenraumes, Annalen der Physik 50 (1916) 385–418.

[74] M. Redhead, Incompleteness, Nonlocality, and Realism: A Prolegomenon to the Philosophy of Quantum Mechanics, Clarendon Press, Oxford, 1990.

[75] I. Pitowsky, Resolution of the Einstein-Podolsky-Rosen and Bell paradoxes, Physical Review Letters 48 (1982) 1299–1302, cf. N. D. Mermin, *Physical Review Letters* **49**, 1214 (1982); A. L. Macdonald, *Physical Review Letters* **49**, 1215 (1982); Itamar Pitowsky, *Physical Review Letters* **49**, 1216 (1982).

[76] D. A. Meyer, Finite precision measurement nullifies the Kochen-Specker theorem, Physical Review Letters 83 (1999) 3751–3754, arXiv:quant-ph/9905080.

[77] I. Kant, Kritik der reinen Vernunft, 2nd Edition, 1787, project Gutenberg. Etext #4280. - First Release: Jul 2003 - ID:4884.

[78] E. Specker, Die Logik nicht gleichzeitig entscheidbarer Aussagen, Dialectica 14 (1960) 175–182, reprinted in [79, pp. 175–182]; English translation: *The logic of propositions which are not simultaneously decidable*, reprinted in [117, pp. 135-140].

[79] E. Specker, Selecta, Birkhäuser Verlag, Basel, 1990.

[80] R. J. Boskovich, De spacio et tempore, ut a nobis cognoscuntur, Vienna, 1755, English translation in [118].

[81] T. Toffoli, The role of the observer in uniform systems, in: G. J. Klir (Ed.), Applied General Systems Research, Recent Developments and Trends, Plenum Press, New York, London, 1978, pp. 395–400.

[82] K. Svozil, On the setting of scales for space and time in arbitrary quantized media, Lawrence Berkeley Laboratory preprint LBL-16097, http://heplibw3.slac.stanford.edu/spires/find/hep?key=1089510 a pdf scan is at
URL http://ccdb3fs.kek.jp/cgi-bin/img/allpdf?198309187

[83] K. Svozil, Connections between deviations from Lorentz transformation and relativistic energy-momentum relation, Europhysics Letters 2 (1986) 83–85, excerpts from [82].

[84] K. Svozil, Operational perception of space-time coordinates in a quantum medium, Il Nuovo Cimento 96B (1986) 127–139.

[85] F. Dyson, Unfashionable pursuits, Mathematical Intelligencer (1983) 47–54Dyson gave related talks; e.g., at Yale, Minnesota and Bonn.

[86] O. E. Rössler, Endophysics, in: J. L. Casti, A. Karlquist (Eds.), Real Brains, Artificial Minds, North-Holland, New York, 1987, p. 25.

[87] O. E. Rössler, Endophysics. Die Welt des inneren Beobachters, Merwe Verlag, Berlin, 1992, with a foreword by Peter Weibel.

[88] K. Svozil, Extrinsic-intrinsec concept and complementarity, in: H. Atmanspacker, G. J. Dalenoort (Eds.), Inside versus Outside, Springer-Verlag, Heidelberg, 1994, pp. 273–288.

[89] O. E. Rössler, Endophysics. The World as an Interface, World Scientific, Singapore, 1998, with a foreword by Peter Weibel.

[90] H. Atmanspacher, G. Dalenoort (Eds.), Inside Versus Outside, Springer, Berlin, 1994.

[91] J. L. Casti, Beyond Believe: Randomness, Prediction and Explanation in Science, CRC Press, Boca Raton, Florida, 1990.

[92] J. L. Casti, Reality Rules. Picturing the World in Mathematics: the Fundamentals, Vol. I, J. Wiley & Sons, New York, 1992.

[93] J. L. Casti, Reality Rules. Picturing the World in Mathematics: the Frontier, Vol. II, J. Wiley & Sons, New York, 1992.

[94] A. Einstein, Zur Elektrodynamik bewegter Körper, Annalen der Physik 17 (1905) 891–921, English translation in the Appendix of [119].

[95] A. D. Alexandrov, On Lorentz transformations, Uspehi Mat. Nauk. 5 (3) (1950) 187.

[96] A. D. Alexandrov, A contribution to chronogeometry, Canadian Journal of Math. 19 (1967) 1119–1128.

[97] A. D. Alexandrov, Mappings of spaces with families of cones and space-time transformations, Annali di Matematica Pura ed Applicata 103 (1967) 229–257.

[98] A. D. Alexandrov, On the principles of relativity theory, in: Classics of Soviet Mathematics. Volume 4. A. D. Alexandrov. Selected Works, 1996, pp. 289–318.

[99] H. J. Borchers, G. C. Hegerfeldt, The structure of space-time transformations, Communications in Mathematical Physics 28 (1972) 259–266.

[100] W. Benz, Geometrische Transformationen, BI Wissenschaftsverlag, Mannheim, 1992.

[101] J. A. Lester, Distance preserving transformations, in: F. Buekenhout (Ed.), Handbook of Incidence Geometry, Elsevier, Amsterdam, 1995.

[102] A. Peres, Defining length, Nature 312 (1984) 10.

[103] J. S. Bell, George Francis FitzGerald, Physics World 5 (1992) 31–35, abridged version by Denis Weaire.

[104] K. Svozil, Relativizing relativity, Foundations of Physics 30 (2000) 1001–1016, e-print arXiv:quant-ph/0001064.

[105] K. Svozil, Conventions in relativity theory and quantum mechanics, Foundations of Physics 32 (2002) 479–502.

[106] K. Svozil, Time generated by intrinsic observers, in: R. Trappl (Ed.), Cybernetics and Systems '96. Proceedings of the 13th European Meeting on Cybernetics and Systems Research, Austrian Society for Cybernetic Studies, Vienna, 1996, pp. 162–166.
URL http://tph.tuwien.ac.at/~svozil/publ/time1.htm

[107] S. Kochen, E. P. Specker, Logical structures arising in quantum theory, in: Symposium on the Theory of Models, Proceedings of the 1963 International Symposium at Berkeley, North Holland, Amsterdam, 1965, pp. 177–189, reprinted in [79, pp. 209–221].

[108] G. 't Hooft, Can quantum mechanics be reconciled with cellular automata?, International Journal of Theoretical Physics 42 (2003) 349–354.
URL http://www.phys.uu.nl/~thooft/gthpub/digit01.ps

[109] M. Navara, V. Rogalewicz, The pasting constructions for orthomodular posets, Mathematische Nachrichten 154 (1991) 157–168.

[110] Cenap S. Ozben, et al, Muon g-2 Collaboration, Precision measurement of the anomalous magnetic moment of the muon.
URL http://arxiv.org/abs/hep-ex/0211044

[111] A. Zeilinger, R. Gähler, C. G. Shull, W. Treimer, W. Mampe, Single- and double-slit diffraction of neutrons, Rev. Mod. Phys. 60 (1988) 1067–1073.
URL http://link.aps.org/abstract/RMP/v60/p1067

[112] K. Zuse, Calculating Space. MIT Technical Translation AZT-70-164-GEMIT, MIT (Proj. MAC), Cambridge, MA, 1970.

[113] H. S. Leff, A. F. Rex, Maxwell's Demon, Princeton University Press, Princeton, 1990.

[114] K. Gödel, in: S. Feferman, J. W. Dawson, Jr., S. C. Kleene, G. H. Moore, R. M. Solovay, J. van Heijenoort (Eds.), Collected Works. Publications 1938-1974. Volume II, Oxford University Press, Oxford, 1990.

[115] R. Landauer, Advertisement for a paper I like, in: J. L. Casti, J. F. Traub (Eds.), On Limits, Santa Fe Institute Report 94-10-056, Santa Fe, NM, 1994, p. 39.

[116] G. J. Chaitin, Information, Randomness and Incompleteness, 2nd Edition, World Scientific, Singapore, 1990, this is a collection of G. Chaitin's publications.

[117] C. A. Hooker, The logico-algebraic approach to quantum mechanics, D. Reidel Pub. Co., Dordrecht; Boston, 1975.

[118] R. J. Boskovich, De spacio et tempore, ut a nobis cognoscuntur, in: J. M. Child (Ed.), A Theory of Natural Philosophy, Open Court (1922) and MIT Press, Cambridge, MA, 1966, pp. 203–205.

[119] A. I. Miller, Albert Einstein's special theory of relativity, Springer, New York, 1998.

Usability of Synchronization for Cognitive Modeling

by Hans H. Diebner and Florian Grond

Abstract

We discuss the synchronization features of a previously introduced adaptive system for dynamics recognition in more detail. We investigate the usability of synchronization for modeling and parameter estimations. It is pointed out inhowfar the adaptive system based on synchronization can become a powerful tool in modeling. The adaptive system can store modules of pre-adapted dynamics and is potentially capable of undergoing self-modification. We compare the stored modules with pre-knowledge that a modeler puts into his or her models. In this sense the adaptive system functions like an expert system.

In recent time the application potential of synchronization mechanisms of dynamical systems has been recognized and increasingly brought up for discussion (cf. e.g. [1]). The applications cover different fields of control [2], parameter estimation [3] and pattern replication [4]. A link of chaos control mechanisms to brain dynamics has been proposed by C. Skarda and W. Freeman [5] and elaborated by others, for a comprehensive review cf. [2].

In this paper we focus on the potentiality of synchronization mechanisms within the fields of life science and cognitive modeling. We point out that synchronization may play an important role in cognitive science and that, therefore, nonlinear science and artificial intelligence studies are closely related. In a recent work [6] we proposed such an adaptive cognitive system that circumvents some of the drawbacks of earlier applications. The most important improvement is a pool of dynamical modules that can be seen as a storage of adapted mirror dynamics which avoid adaptations from scratch. This accounts for two important features of cognitive systems namely the simulation or mirroring capability as discussed in [7, 8] and the usage of *a priori* information.

The paper in hand discusses open problems we encountered in the aforementioned work. Specifically we investigate the efficacy of synchronization with respect to the chosen mode of coupling and the dynamical behavior. We therefore briefly recapitulate the underlying basic mechanism. Assume $\mathbf{x}$ and $\mathbf{x}'$ to be the states of two dynamical systems of the same dimension d and the same dynamics $\mathbf{f}$ which are given by the differential equations

$$\begin{aligned} \dot{\mathbf{x}} &= \mathbf{f}(\mathbf{x};\beta), & \beta &= (\beta_1,\beta_2,\ldots,\beta_m) \\ \dot{\mathbf{x}}' &= \mathbf{f}(\mathbf{x}';\beta'), & \beta' &= (\beta'_1,\beta'_2,\ldots,\beta'_m), \end{aligned} \qquad (1)$$

where β and β' are sets of m fixed parameters. Assume further that at least one of the parameters of the unprimed system is different from the corresponding one in the primed system, $\beta_k \neq \beta'_k$ (for at least one k). If now the difference of at least one pair of corresponding variables multiplied by a factor K_i is added to the unprimed system,

$$\begin{aligned}
\dot{x}_1 &= f_1(x_1, x_2, \ldots, x_n; \beta) + K_1(x'_1 - x_1) \\
\dot{x}_2 &= f_2(x_1, x_2, \ldots, x_n; \beta) + K_2(x'_2 - x_2) \\
&\vdots \\
\dot{x}_n &= f_n(x_1, x_2, \ldots, x_n; \beta) + K_n(x'_n - x_n)
\end{aligned} \qquad (2)$$

$$K_i \geq 0 \quad (K_j > 0 \text{ for at least one } j \in \{1, \ldots, d\}),$$

this system will be forced to the dynamics of the primed controlling system. The success of this forcing depends on coupling modes and dynamical features as investigated in the sequel.

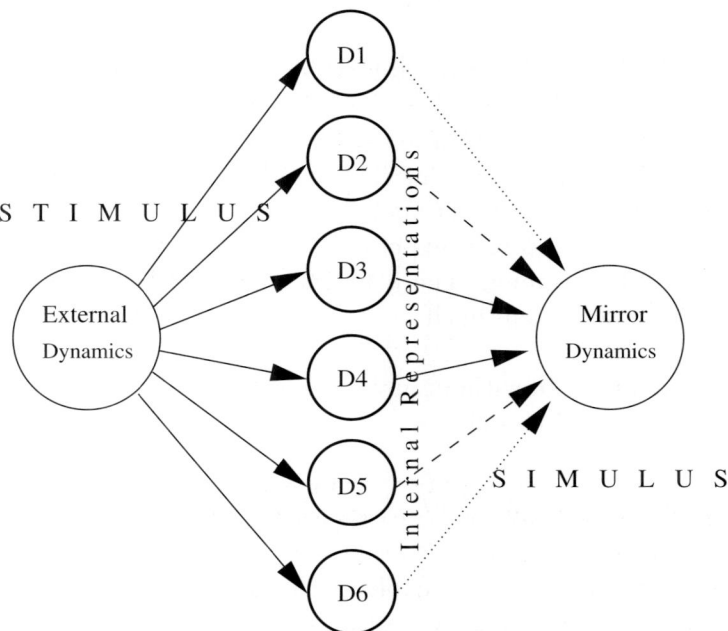

Figure 1: Cognitive system that adapts a simulus to a 'perceived' stimulus using pre-adapted modules for an efficient adaptation.

Before we proceed with this investigation we briefly outline the scheme of the adaptive system which is depicted in Fig. 1. An external system – experienced through a stimulus – forces internal modules each according to Eq. 2. One of the modules which we call simulus undergoes an adaptation to the external system. The other modules constitute a set of dynamics that more or less resemble the external one and are compared to it. The resemblance is supposed to be reflected in the magnitude of the forcing. The information on the difference or resemblance is used in a superpositional manner to construct a simulus.

For example, assume given a Rossler system as external dynamics represented as time series of at least one variable x'_i then the modules may be defined through a set of Rossler systems differing in the value of a certain parameter. Assume the value of the corresponding external parameter to lie within the span of the internal parameter values. The idea now is to vary the parameter value of the simulus until the forcing term $x'_i - x_i^{\text{simulus}}$ vanishes (for all used components i) which indicates adaptation. This adaptation can in principle been done without using the modules. The magnitudes of the forcing terms $x'_i - x_i^{\text{module}_j}$ of the modules, however, supply additional information and are used as storage of previously experienced dynamics.

Basic parameter estimation procedures have been described in [9, 10] where *a priori* knowledge on the type of external dynamics was used. The usage of modules that span a grid around the external dynamics in our model considerably enhanced the performance of the adaptive system. Additionally, the model allows to abandon *a priori* knowledge to a certain degree. The modules play the role of previously experienced dynamics that allow a quicker re-adaptation and fine tuning. The *a priori* knowledge, so to say, is part of the artificial cognitive system itself.

To start the adaptation without any *a priori* knowledge one may formulate a general n−dimensional model

$$\dot{x}_i = \beta_{i00} + \sum_{j=1}^{n} \beta_{ij0} x_j + \sum_{j=1}^{n}\sum_{k=1}^{n} \beta_{ijk} x_j x_k + \ldots, \qquad (3)$$

up to a necessary but still manageable order (according to Occam's razor). Since the adaptation performance depends on the number of free parameters one can increasingly benefit from pre-adapted modules in the course of time, i.e. from stored *a priori* knowledge. We mention in passing that a self-modifying feature can be introduced through forcing between the internal modules in combination with appropriate evolutionary criteria of fitness. We expect further improvement with respect to an evolutionary optimization (cf. [11]).

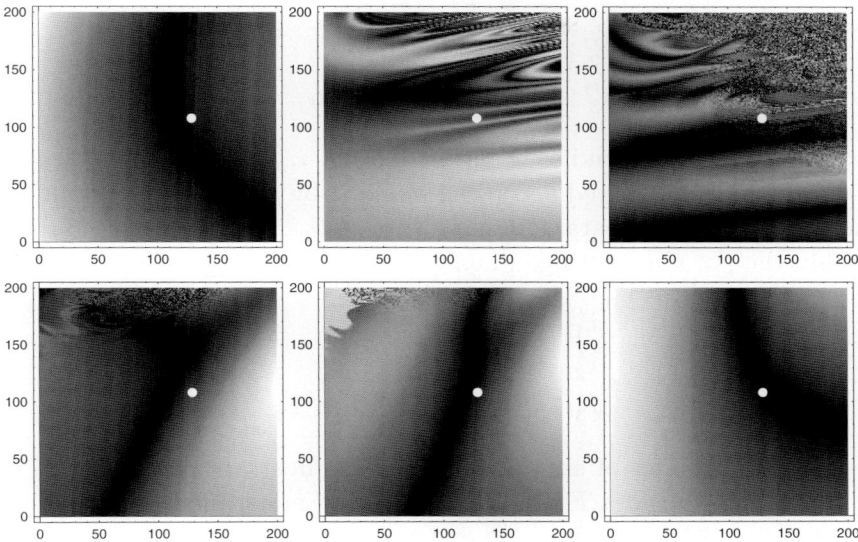

Figure 2: Synchronization behavior of Rossler's system. The plane of each screenshot shows the increments of the parameter space that define an array of the 200×200 controlled systems. The white circle marks the pair of parameters $(\alpha' = 5.7, \beta' = 0.2)$ that defines the control system. The differences in the first variable between control and controlled systems is depicted as grey code. Cf. text for a detailed description.

At a first glance it is obvious to use the magnitude of the forcing terms $F_i = |x'_i - x_i^{\text{simulus}}|$ as a criterion for the goodness of synchronization or adaptation, respectively. This surely holds if the F_i almost vanish at every time instant. However, a larger F_i does not necessarily imply that the module does not resemble the target system well due to the fact that some dynamical types are extremely hard to synchronize although they are quite similar. Depending in the mode of forcing even identical dynamics differing only in the intitial state may be hard to synchronize. For example, we observed that two Lorenz systems require a relatively large forcing strength K_i in a certain parameter range in order to be synchronized even if the two systems differ only in the initial states. This behavior has to be taken into account when constructing an adaptive system using synchronization mechanisms. In the following we explore this behavior in more detail.

We start by investigating the Rossler system. Assume given a set of modules defined by

$$\begin{aligned}
\dot{x}_1^{ij} &= -x_2^{ij} - x_3^{ij} \\
\dot{x}_2^{ij} &= x_1^{ij} + \beta_i x_2^{ij} \\
\dot{x}_3^{ij} &= 0.2 + x_1^{ij} x_3^{ij} - \alpha_j x_3^{ij},
\end{aligned} \quad (4)$$

with $\beta_i = 0.02 + 0.0014i$ and $\alpha_j = 3.0 + 0.025j$; $i, j = 0, \ldots, n-1$; $n = 200$. All $n \times n$ modules are forced according to Eq. 2 by an external system defined through a fixed pair of parameters $\beta_0 \leq \beta' \leq \beta_{n-1}$, $\alpha_0 \leq \alpha' \leq \alpha_{n-1}$) that lies within the parameter range, i.e.,

$$\begin{aligned}
\dot{x}_1' &= -x_2' - x_3' \\
\dot{x}_2' &= x_1' + \beta' x_2' \\
\dot{x}_3' &= 0.2 + x_1' x_3' - \alpha' x_3'.
\end{aligned} \quad (5)$$

Figure 2 shows a series of six screenshots that illustrate the evolution of the $n \times n$ forcing magnitudes $F_{ij} = |x_1' - x_1^{ij}|$. The magnitudes F_{ij} are depicted as grey code on the $\{i, j\}$-plane, whereby the grey code has been re-scaled for each screenshot. Black areas code for almost vanishing $F_{ij}s$ and white areas indicate large differences between external system and module. The parameter of the external system is marked as a filled white circle.

The simulation has been started with zero forcing ($K_i = 0$ for all three components i) but with identical intial states of all modules and the external system. Therefore, all $F_{ij} = 0$ for $t = 0$. The first screenshot depicts the $F_{ij}s$ after a small time duration $t = 148$ (iteration steps) where emerging differences can be observed indicated through the bright areas. After a further time duration (at $t = 1433$) the chaotic regime can be clearly recognized by a heavily fluctuating area emerging in the upper right corner indicating the absence of phase correlations between the neighbored modules. In other areas, however, one observes standing wave-like oscillations indicating a constant phase relation of the neighbored modules but with different amplitudes of the oscillations. The "iso-phases" can be clearly recognized and distinguished from the jittered uncorrelated area. This feature is more pronounced after a further time duration at $t = 4120$ seen in the third screenshot. One can distinct areas of horizontal from areas of curved iso-phases suggesting periodic regimes of similar frequencies that lie outside and, in form of "islands", inside the chaotic regime, respectively.

The three screenshots of the second row show the behavior of the F_{ij}s after setting $K_1 = 0.5$ at $t = 4120$. The leftmost screenshot shows an emerging synchronization at $t = 6969$. A remaining small area of non-correlated phases can be observed in the upper part. Please note, that the grey code has been re-scaled. Otherwise a homogeneous grey value would indicate lower differences in the mean compared to the beginning hindering to recognize spatial inhomogeneous behavior which we want to emphasize here. Screenshots five and six have been taken at $t = 9954$ and $t = 17364$, respectively. The fifth image unmasks the upper left corner as a synchronization resistant area in the given forcing mode. Therefore K_1 has been increased to $K_1 = 0.87$ at $t = 9954$ which leads to a stronger synchronization shown in the last image.

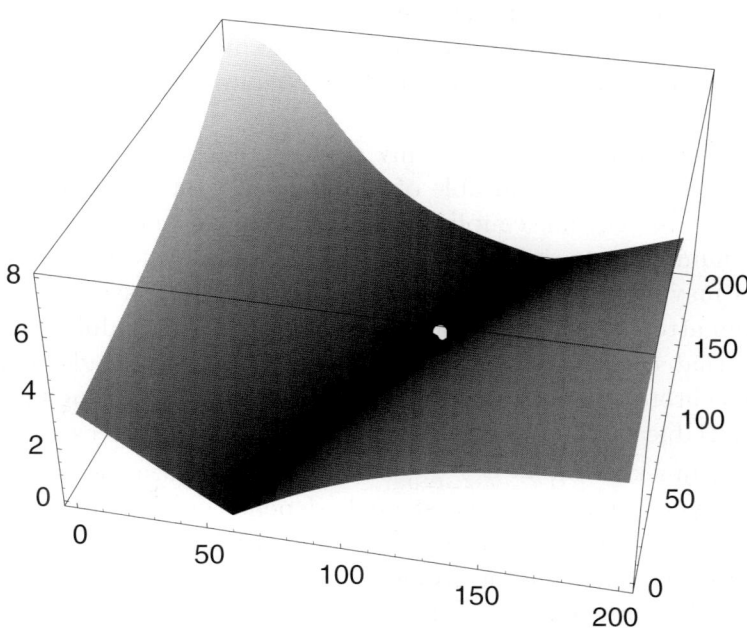

Figure 3: Cumulated forcing terms shown as a surface function over the parameter space. The defining parameter pair of the control system is shown as a white mark. Cf. text for a detailed description.

The "live" visualization reveals some features that are hard to capture with static screenshots. The dynamical behavior of the F_{ij}s shows a characteristic spiral-like wave front moving around the point in the parameter plane that marks the external system. Because of this waxing and waning behavior there is a remaining uncertainty when taking the F_{ij}s as an instantaneous measure for the difference of dynamical systems. However, if one cumulates the F_{ij}s over a time interval, T, according to[23]

$$S_{ij} = \frac{1}{T} \sum_{t=0}^{T-1} |x'_1 - x_1^{ij}|, \qquad (6)$$

one observes an emerging stationary "potential" surface S_{ij} as depicted in Fig. 3. The function S_{ij} can be interpreted as a likelihood-based function that is to be minimized with respect to the free parameters as pointed out in [9]. If the systems do not synchronize the cumulated forcing term of Eq. 6 leads simply to the average distance between the states of control and un-controlled system. With respect to the momentarily given differences F_{ij} the resulting "potential surface" S_{ij} is stationary and smoothed. However, as can be clearly seen in Fig. 3 the potential surface can lead to unpronounced minima with respect to particular parameters, β in the given example.

In a realistic situation one has only a scalar time series available which is expected to represent one variable of a dynamical system. This is the main reason why we use only one coupling variable although we would have all three. The exact number of variables is usually not even known and the measured time series may be a lumped variable of the underlying system. What can be seen from attempts to synchronize the Rossler system in different variables is the fact that coupling in the first and the second one leads to relatively accurate synchronization results whereas synchronization through coupling in the third variable fails. However, we experienced in similar cases that using the modified function $S_{ij} = \frac{1}{T} \sum_{t=0}^{T-1} e^{|x'_1 - x_1^{ij}|}$ synchronization is possible. In general, the synchronizability also strongly depends on the number of free parameters and many other settings. This indicates that experienced modelers are often capable of tackling the problem nevertheless, for example by reducing the number of parameters or formulating a modified likelihood as mentioned. In this context we want to emphasize a possible support by a cognitive system. The stored pre-adapted modules of the adaptive system may serve as a kind of expert system that amends the experience of the modeler [12].

The following example demonstrates that it is possible to start the adaptive procedure with little *a priori* knowledge and without pre-adapted modules. Therefore, we use the maximum likelihood approach to generate a first module by applying it to a modified Rossler system that is relatively hard to synchronize due to a more complex dynamical behavior of the switching

---

[23] We restrict the coupling to the first component.

variable as can be seen in the left part of Fig. 4. We use the first variable x_1 of

$$\begin{aligned}
\dot{x}_1' &= -0.97x_2' - 0.97x_3' \\
\dot{x}_2' &= x_1' + 0.19x_2' - 1.024x_3' \\
\dot{x}_3' &= 0.2 + x_1'x_3' - 5x_3'.
\end{aligned} \quad (7)$$

as an "observed" time series produced with an integration time step of $\Delta t = 0.08$ and total lenght of $T = 10000$ steps. As model system that is to be adapted to the time series we choose

$$\begin{aligned}
\dot{x}_1 &= \beta_1 + \beta_2 x_1 + \beta_3 x_2 + \beta_4 x_3 \\
&+ \beta_5 x_1 x_2 + \beta_6 x_1 x_3 + \beta_7 x_2 x_3 \\
\dot{x}_2 &= \beta_8 + \beta_9 x_1 + \beta_{10} x_2 + \beta_{11} x_3 \\
&+ \beta_{12} x_1 x_2 + \beta_{13} x_1 x_3 + \beta_{14} x_2 x_3 \\
\dot{x}_3 &= \beta_{15} + \beta_{16} x_1 + \beta_{17} x_2 + \beta_{18} x_3 \\
&+ \beta_{19} x_1 x_2 + \beta_{20} x_1 x_3 + \beta_{21} x_2 x_3
\end{aligned} \quad (8)$$

without any a priori information on the 21 free parameters. In other words, we start with zero initial values for all 21 parameters. The initial values of the model variables have been set to the "observed" values. The forcing strenght is chosen to be $K_1 = 0.5$. The minimization of $S = \frac{1}{T}\sum_{t=0}^{9999}|x_1'(t) - x_1(t)|$, is executed by using a standard Nelder-Mead minimization procedure. The model output has been computed with a fourth order Runge-Kutta algorithm using a time step of $\Delta t = 0.08$ to mimick the same sampling frequency as the "measured" time series. We repeated the minimization, as usual in minimization procedures with multi-parameter functions, with randomly chosen intitial guesses for the parameter values. We stopped the procedure after 100000 runs that did not lead to further improvement of the likelihood.

The optimization resulted in the following set of parameter values:

$$\begin{aligned}
\beta_1,\ldots,\beta_7 &= 0.90,\ 1.15,\ -1.92,\ -2.31,\ \ 0.00,\ \ \ 0.00,\ \ 0.00 \\
\beta_8,\ldots,\beta_{14} &= 0.31,\ 0.86,\ -0.92,\ -0.26,\ \ 0.07,\ -0.29,\ \ 0.00 \\
\beta_{15},\ldots,\beta_{21} &= 0.29,\ 0.18,\ \ \ 0.94,\ -5.05,\ -0.21,\ \ 0.93,\ \ 0.01.
\end{aligned} \quad (9)$$

The right part of Figure 4 shows the estimated attractor. It remains to be shown that the estimated attractor is affine to the original one.

To summarize, we presented evidence that synchronization has the potentiality to become a powerful method in the fields of dynamics recognition. The success of modeling by means of synchronization quite freqently depends on subtle modifications by an experienced modeler. There is a lack of both a clear

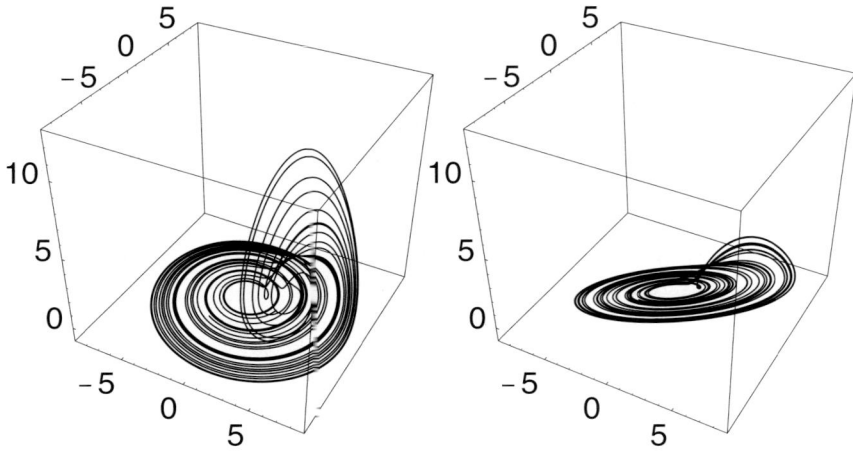

Figure 4: Phase space of the modified Rossler system of Eq. 7 (left part) and the estimated attractor (right part).

criterion which mode of application of the synchronization procedure will be successful and an absolute criterion of success. For example, in a noisy or distorted time series the synchronization may be "successful" with respect to the distortion. This has to be tested from case to case. We argue that to unfold the power of synchronization new approaches are necessary to tackle this lack of full mathematical analyticity. We investigated a recently proposed cognitive system under this aspect where synchronization reflects a basic mimetic feature in a more general sense as pure parameter estimation. The main idea behind this adaptive system is to mimick to some degree the experience-based *modus operandi* of experts by bringing in *a priori* knowledge. We also demand for an opening of scientific principles to a more performative approach that includes simulations and visualizations as a serious tool to gain knowledge. Since the process of understanding obviously takes place in a highly nonlinear system we think that artificial intelligence research and nonlinear science should go hand in hand.

Acknowledgments

We dedicate this article to our friend Mohamed El Naschie on the occasion of his 60th birthday and express our deep respect towards his unique openness and generosity.

References

[1] Erik Mosekilde, Yuri Maistrenko, and Dmitry Postnov, editors. *Chaotic Synchronization: Application to Living Systems*. World Scientific, Singapore, 2002.

[2] Axel A. Hoff. *Chaoskontrolle, Informationsverarbeitung und chemische Reaktionssysteme*. Logos Verlag, Berlin, 1997.

[3] Holger Kantz and Thomas Schreiber. *Nonlinear Time Series Analysis*. Cambridge University Press, Cambridge, 1997.

[4] Vladimir I. Nekorkin and Manuel G. Velarde. *Synergetic Phenomena in Active Lattices, Patterns, Waves, Solitons, Chaos*. Springer, Berlin, 2002.

[5] C. Skarda and W.J. Freeman. How brains make chaos in order to make sense of the world. *Behav. Brain Sci.*, 10:161–195, 1987.

[6] Hans H. Diebner, Axel A. Hoff, Adolf Mathias, Horst Prehn, Marco Rohrbach, and Sven Sahle. Control and adaptation of spatio-temporal patterns. *Z. Naturforsch. A*, 56:663–669, 2001.

[7] V. Gallese, L. Fadiga, and G. Rizzolatti. Action recognition in the premotor cortex. *Brain*, 119:593–609, 1996.

[8] G. Rizzolatti, L. Fadiga, V. Gallese, and L. Fogassi. Premotor cortex and the recognition of motor actions. *Cognitive Brain Research*, 3:131–141, 1996.

[9] U. Parlitz. Estimating model parameters from time series by autosynchronization. *Phys. Rev. Lett.*, 76:1232, 1996.

[10] A. Maybhate and R. E. Amritkar. Dynamic algorithm for parameter estimation and its applications. *Physical Review E*, 61:6461–6470, 2000.

[11] G. Kampis. *Self-Modifying Systems in Biology and Cognitive Sciences*. Pergamon Press, Oxford, 1991.

[12] Hans H. Diebner. Operational hermeneutics and communication. In Hans H. Diebner and Lehan Ramsay, editors, *Hierarchies of Communication*. ZKM edition, Karlsruhe, 2003.

Riddling Bifurcation and ... Interstellar Journeys

by Tomasz Kapitaniak

Abstract

We show that riddling bifurcation which is characteristic for low-dimensional attractors embedded in higher-dimensional phase space can give physical mechanism explaining interstellar journeys described in science-fiction literature.

Phase spaces of many dynamical systems describing nature are higher- or even infinite-dimensional ones. The classical examples are systems described by a ordinary differential equation. However, in many cases observable phenomena of practical importance occur on (or in the neighborhood) of low-dimensional manifold [1]. In this paper we argue that the transition from lower-dimensional attractor to the higher-dimensional one can explain the way in which interstellar journeys are described in science-fiction literature.

Consider two identical chaotic systems $x_{n+1} = f(x_n)$ and $y_{n+1} = f(y_n)$, $x, y \in R$ evolving on an asymptotically stable chaotic attractor A coupled as

$$x_{n+1} = f(x_n) + d_1(y_n - x_n),$$

$$y_{n+1} = f(y_n) + d_2(x_n - y_n). \qquad (1)$$

It is well-known that this systems can synchronize for some ranges of $d_{1,2} \in R$, i.e., $|x_n - y_n| \to 0$ as $n \to \infty$ [6]. In the complete synchronized regime, the dynamics of the coupled system (1) is restricted to one-dimensional invariant subspace $x_n = y_n$, so we have the classical example of the system which is 2-dimensional but the dynamics takes place on the one-dimensional attractor A.

The infinite number of unstable periodic orbits (UPOs) is embedded in a chaotic attractor A [2]. UPOs provide the skeleton of the attractor and it allows for the characterization and the estimation in a fundamental way of many dynamical invariants. UPOs play a fundamental role in the mechanism of destabilization of the chaotic attrator localized in some symmetric invariant manifold and it is responsible for the dynamics of the phenomena such as riddling of the basin of attraction and bubbling of the chaotic attractor [5]. Recently, UPOs have also been used in the description of higher-dimensional dynamical phenomena of chaos-hyperchaos transition (i.e., transition from the attractor characterized by one positive Lyapunov exponent to the attractor characterized by at least two positive exponents) [3].

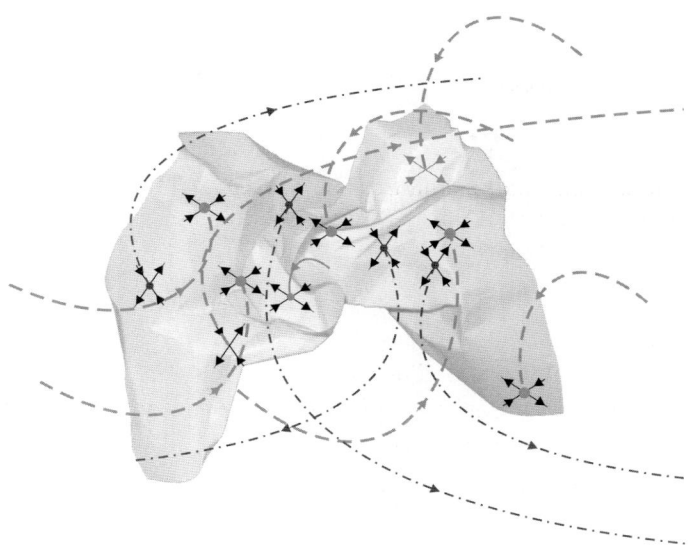

Figure 1: Chaotic attractor A with UPO with stable (dash) and unstable (dash-dot) transverse manifolds

The simultaneous existence of UPOs with different number of unstable direction gives rise to a new kind of nonhyperbolicity known as unstable dimension variability [4] and it may give a possible dynamic mechanism for interstellar journeys.

Consider the n-dimensional chaotic attractor A located in a m-dimensional phase space ($n < m$), as shown in Fig. 1. Assume that UPOs emmbedded in the attractor have already undergone riddling bifurcation [8] so in the attractor two types of UPOs are embedded. The first type are the orbits with stable transverse manifold (indicated in thick, gray dashed lines in Fig. 1) and the second type these with unstable transverse manifold (indicated as thin, dash-dot, black lines). Suppose that the phase space trajectory on the attractor A is close to the point a and that we have to implement the control procedure which allows us to go from the point a to the point $b \in A$ (Fig. 2). The straightforward way is to restrict the path from a to b to the attractor A (pieces of very thick, gray lines in Fig. 2). Due to the ergodicity point b will be reached in finite time but this could be too long for practical acceptance. The alternative approach is the path in the neighborhood of the attractor A like the thick, black, solid line in Fig. 2. In this case the phase space trajectory has to leave the attractor A, stay in its neighborhood and return to the

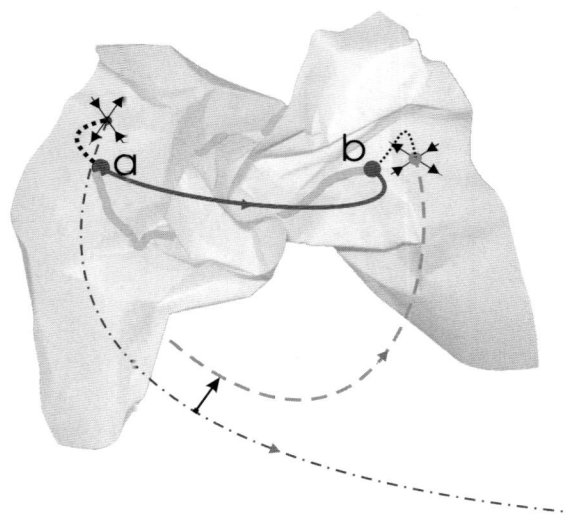

Figure 2: Controlling procedure

attractor in the appropriate point. Dynamically such a situation is possible when in the neighborhood of the point a there exists UPO with unstable transverse manifold then by applying control one can move the trajectory to the neighborhood of this UPO (along the thick, dotted line in Fig. 2) and allow it to leave the attractor (along the thin, dash-dot, black line in Fig. 2). After leaving the attractor the trajectory has to come close to the stable manifold of the other UPO with stable transverse manifold (thick, dashed line in Fig. 2). At this point the control moving the trajectory to the stable manifold has to be applied. Along the dashed manifold the trajectory returns to the attractor in the neighborhood of the point b. After another application of the control the trajectory reaches the target point b (thin, dotted line line in Fig. 2).

Now recall one of the fundamental problems in the discussion of the possible interstellar journeys. Due to the huge distance in the universe and speed limitations of the space crafts, the journeys to the other stellar systems seem to be impossible (at least at the current state of science and technology) carried or even imagine. On the other hand, there are plenty of descriptions of these journeys in science-fiction literature (for example [7]). In the great number of novels such journeys are possible due to the existence of super- (or extra-) space which exists besides the universe in which human beings are living. Huge distances are covered by entering this super-space and returning to the home universe in the appropriate point. Usually there are some limitations: (i) super-space is not reachable from any point of the universe but only from the neighborhood of some special points, (ii) the energy is necessary for entering and leaving super-space, (iii)the return to the target point of the home universe is not always possible but the return in the neighborhood of the target is possible.

Now assume that (i) the points which allow entering super-space are UPOs with transversally unstable manifolds, (ii) the return to the universe is possible along the transversally stable manifolds of the appropriate UPO, (iii) the energy is necessary to reach the neighborhood of appropriate UPO and to change manifolds in the super-space.

At this point, one immediately finds out the analogy between the controlling procedure based on Fig. 2 and interstellar journeys (at least these described in science-fiction literature). The existence of the points in the universe which underwent riddling bifurcation (black holes,....??) is the necessary condition for interstellar journeys.

References

[1] D. Ruelle, F. Takens. Commun Math. phys. 1971; 20: 167, W.L. Ditto *et al*. Phys. Rev. Lett. 1989; 63: 923, W.L. Ditto, S.N. Rauseo, M.L. Spano. Phys. Rev. Lett. 1990;65:3211, Y.-C.Lai, E.M. Bolt, Z. Liu. Chaos, Solitons and Fractals 2003;15:219

[2] D. Auerbach, P. Cvitanovic, J.-P. Eckmann, G.H. Gunaratne, and I. Procaccia, Phys. Rev. Lett. **58**, 2387; G.H. Gunaratne and I. Procaccia, *ibid.*, **59**, 1377 (1987).

[3] O.E. Rossler, Phys. Lett. A **57**, 397 (1976); Z. Naturforsch, A**31**, 1168 (1976), *ibid* **38** 788 (1979), J. Peinke, J. Parisi, O.E. Rossler and R. Stoop, *Encounter with chaos*, (Springer: Berlin 1992), G. Baier and M. Klein, *A Chaotic Hierarchy*, (World Scientific, Singapore, 1991), T. Kapitaniak, L.O. Chua, and G.-Q. Zhong, IEEE Trans. CAS, **41**, 499 (1994), T. Kapitaniak and L.O. Chua, Int. J. Bifur. Chaos, **4**, 477 (1994), T. Kapitaniak and W.-H. Steeb, Phys. Lett. A **152**, 33 (1991), T. Kapitaniak, Phys. Rev. E **47**, R2975 (1993), M.A. Harrison and Y.-Ch. Lai, Phys. Rev E **59**, R3799 (1999), K. Stefanski, Chaos, Solitons Francats **9**, 83 (1998), T. Kapitaniak, Yu. Maistrenko, and S. Popovich, Phys. Rev. E **62**, 1972 (2000).

[4] R. Abraham and S. Smale, Proc. Symp. Pure Math. (AMS) **14**, 5 (1970); E.J. Kostelich, I. Kan, C. Grebogi, E. Ott, and J.A. Yorke, Physica D **109**, 81 (1997); Y.-Ch. Lai, C. Grebogi, and J.Kurths, Phys. Rev. E 59, 2907 (1999); Y.-Ch. Lai, *ibid* **59**, R3803 (1999); Y.-Ch. Lai and C. Grebogi, Phys. Rev. Lett., **83**, 2926 (1999); R.L. Viana and C. Grebogi, *ibid* **62**, 462 (2000); R.L. Viana and C. Grebogi, Intern. J. Bif and Chaos, **11**, 2689 (2001); E.E.N. Macau, C. Grebogi, and Y.-Ch. Lai, Phys. Rev. E**65**, 027202 (2002).

[5] Y. Nagai, and Y.-Ch. Lai, Phys. Rev. E **55**, R1251 (1997); Y.-Ch. Lai, Phys. Rev. E **56**, 1407 (1997); Y. Nagai and Y.-Ch. Lai, Phys. Rev. E **56**, 4031 (1997), Y.-Ch. Lai, Phys. Rev E **59**, R3803, (1999)

[6] H. Fuijsaka, and T. Yamada, Prog.Theor.Phys. 70, 1240 (1983), V.S Afraimovich, N.N. Verichev, and M.I. Rabinovich, Radiophysics and Quantum Electronics 29, 795 (1986), L. Pecora, and T.S. Carroll, Phys.Rev.Lett. 64, 821 (1990), M. De Sousa, A.J. Lichtenberg, and M.A. Lieberman, Phys. Rev. A46, 7359 (1992), Y.-C. Lai, and C. Grebogi, Phys.Rev. 47E., 2357 (1993), T. Kapitaniak, Physical Review, 50E, 1642 (1994), Yu. Maistrenko, and T. Kapitaniak, Phys. Rev., 54E, 3285 (1996)

[7] I. Asimov, Pirates of the asteroids, New English Library, Holborn (1974)

[8] Y.-Ch. Lai, C. Grebogi, J.A. Yorke, and S.C. Venkataramani, Phys. Rev. Lett. **77**, 55 (1996), V. Astakhov, A. Shabunin, T. Kapitaniak, and V. Anishchenko, *ibid*, **79**, 1014 (1997), T. Kapitaniak, Yu. Maistrenko and C. Grebogi, Chaos, Solitons and Fractals, 17, 61 (2003).

Theoretical Poroelasticity – A New Approach

by Reint de Boer

Abstract

In this paper a consistent poroelasticity for a ternary mathematical model, consisting of a compressible porous solid filled with a mixture of an incompressible liquid and a compressible gas, is created. On the basis of the Theory of Porous Media (TPM) the fundamental main equations of poroelasticity are worked out, namely the equations of motion expressed by the displacement vectors.

1 Introduction

Experience teaches us that solids under the action of external forces show a change in their shapes which vanishes after a gradual unloading or caused vibration of the solid by rapid unloading. This material behavior is known as an elastic phenomenon and is probably the most investigated mechanical problem.

Already in the first stages of the development of mechanics in the 17th and 18th century elastic problems attracted attention, e.g. by Jakob Bernoulli who investigated – obviously for the first time – the bending of an elastic beam and Leonard Euler who investigated the so-called elastica problems. These investigations, were primarily based on ad-hoc assumptions – the stress concept and the constitutive equation were not known – and were concerned with the determination of the elastic curves of thin beams at finite displacement, however with small strains. In this connection Euler succeeded in developing a formula for the critical load of a slender rod.

Because of the lack of the stress concept and a constitutive equation for elastic materials as aforementioned the scientists in the 17th and 18th century were not able to create a continuummechanical elasticy theory. This happened in the first half of the 19th century. It was the ingenious mathematician Augustin-Louis Cauchy who created the stress concept and later stated what is today known as Cauchy's first and second laws of motion (see de Boer, 2000)[24], namely the principles of linear momentum and moment of momentum. The analysis of stress concept necessitates, the analysis of the deformation concept, which flows into the concept of the strain tensor. Cauchy first developed this concept for the arbitrary displacements u, v, and w which, like the stress tensor, he related to the deformed system. Cauchy later went

---

[24]Names connected with year dates point to the references at the end of this article.

2 Material-Independent Relations

2.1 Introductory Remarks on Porous Media Theory

In many branches of engineering, for example, in chemical engineering, material science, and soil mechanics, as well as in biomechanics, the different reactions of material systems undergoing external and/or internal loadings must be studied and described precisely in order to be able to predict the responses of these systems. Subsequently, the most important point of the investigation is to determine the composition of the body first, because one must know the physically and chemically differing materials that constitute the system under consideration. The material systems (or bodies) in these fields of engineering can be composed in various ways. On the one hand, solids can consist of different solid components, such as dense concrete, without considerable pores. On the other hand, solids can contain closed and open pores, such as ceramics and soils, as well as concrete. The pores can be filled with fluids and, due to the different material properties and the different motions, there may be interaction between the constituents. This fact makes the description of the mechanical (or the thermomechanical) behavior difficult.

Because the exact description of the location of the pores (empty or filled with fluids) in the solid material is nearly impossible, a heterogeneous composition can be investigated by using the volume fraction concept (see next section). This concept results in homogenized ("smeared") continua with reduced densities for the solid and fluid phases arise which can then be treated by the mixture theory.

The combination of the mixture theory with the volume fraction concept touches the microscale. The question arises as to what scale the mechanical or thermodynamic investigations should be performed, on the macro- or microscale? In principle, both strategies are possible. However, the micromechanical approach, with all its averaging processes, leads to a huge formalism (see, e.g., de Boer and Didwania, 1997), revealing some important mechanical relations. Because of the fact that this approach is still in its infancy, only the macroscopic theory, which has only recently come to consistent conclusions, will be discussed in the following pages. Of course, the micromechanical effects which are raised by the volume fraction concept will be considered using macromechanical quantities.

Another major problem arises in macroscopic porous media concerning the closure problem. It can easily be shown that, for porous media consisting of α constituents, $(\alpha - 1)$ field equations are missing. However, this is only a result of the macroscopic point of view. Nevertheless, as was mentioned, the porous media theory touches the microscale due to the volume fraction concept. Thus, concerning the closure problem, the microscale should be taken into consideration. Following this idea, the discussion on the closure problem results in some very reasonable conclusions.

The aim of the next sections is to develop a consistent macroscopic porous media theory for a mathematical model with compressible elastic, capillary-porous solid and compressible gas phases and an incompressible liquid which would contain the results of the former macroscopic theories as special cases.

2.2 Volume fraction concept

In the volume fraction concept, it is assumed that the porous solid always models a control space and that only the liquids and/or gases contained in the pores can leave the control space. Furthermore, it is assumed that the pores are statistically distributed and that an arbitrary volume element is composed of the volume elements of the real constituents. These volume elements dv^α ($\alpha = S, L$ and G are the solid, liquid and gas phases, respectively) are related to the volume dv of the control space by the volume fraction n^α. Thus with

$$dv^\alpha = n^\alpha dv \quad (2.1)$$

homogenized solid and liquid or gas constituents are created with reduced densities

$$\rho^\alpha = n^\alpha \rho^{\alpha R}, \quad (2.2)$$

where $\rho^{\alpha R}$ is the real density of the constituent ρ^α.
The volume fractions in (2.1) and (2.2) satisfy the volume fraction condition (or saturation)

$$\sum_\alpha n^\alpha = 1. \quad (2.3)$$

This condition has turned out to be an important constraint in the constitutive theory of porous media.

2.3 Kinematics

The kinematics in the porous media theory are based on two fundamental assumptions:

(1) Each spatial point $\mathbf{x}$ of the actual placement is simultaneously occupied by material points X_α of all κ constituents φ^α at the time t. The material points proceed from different reference positions $\mathbf{X}_\alpha$ at time $t = t_0$.

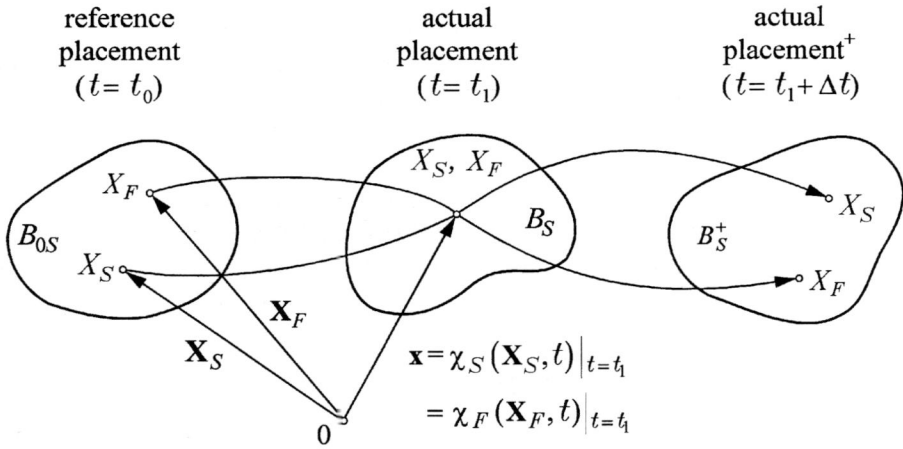

Figure 1: Illustration of the motion of a solid and a liquid particle in a liquid saturated porous solid.

(2) Each constituent is assigned an independent state of motion.

It should, however, be mentioned that the volumetric strains are constrained by the saturation condition. If the motion of the constituent is understood as a chronological succession of placements χ_α, then for the spatial position vector $\mathbf{x}$ of the material points X_α, which can be identified with the reference position vector $\mathbf{X}_\alpha$ at time $t = t_0$, the following relation holds at time t:

$$\mathbf{x} = \chi_\alpha(\mathbf{X}_\alpha, t) \, . \quad (2.4)$$

The position vector $\mathbf{x}$ is an element of the control space of the porous solid at time t. In general, it is not necessary to demand that the reference positions $\mathbf{X}_L$ and $\mathbf{X}_G$ of the liquid and gas particles are elements of the reference placement of the solid phase at time $t = t_0$, i.e., $\mathbf{X}_L \notin B_{0S}$ and $\mathbf{X}_G \notin B_{0S}$. Only for such deformation processes in which the fluid phases leave the control space of the solid phase are the reference positions $\mathbf{X}_L$ and $\mathbf{X}_G$ elements of B_{0S} (see Bluhm, 1997). A geometrical interpretation of the motion function (2.4), concerning the motion of a solid and a fluid particle, is shown in Fig. 1.

Eqn. (2.4) represents the Lagrange description of motion. The function χ_α is postulated to be unique, and uniquely invertible, at any time t. The existence of a function inverse to (2.4) leads to the Eulerian description of motion, viz.

$$\mathbf{X}_\alpha = \boldsymbol{\chi}_\alpha^{-1}(\mathbf{x}, t) . \quad (2.5)$$

A mathematically necessary and sufficient condition for the existence of Eqn. (2.5) is given, if the Jacobian

$$J_\alpha = \det \mathbf{F}_\alpha \quad (2.6)$$

differs from zero. In (2.6), $\mathbf{F}_\alpha$ is the deformation gradient, which is defined as

$$\mathbf{F}_\alpha = \mathrm{Grad}_\alpha \, \boldsymbol{\chi}_\alpha . \quad (2.7)$$

The differential operator "Grad_α" denotes a partial differentiation with respect to the reference position $\mathbf{X}_\alpha$ of the constituents φ^α. The inverse of (2.7) is given by

$$\mathbf{F}_\alpha^{-1} = \mathrm{grad}\, \mathbf{X}_\alpha , \quad (2.8)$$

with the differential operator "grad" referring to the spatial point $\mathbf{x}$. During the deformation process, $\mathbf{F}_\alpha$ is restricted to

$$\det \mathbf{F}_\alpha > 0 . \quad (2.9)$$

With the Lagrange description of the motion (2.4), the velocity and the acceleration of a material point of a constituent φ^α are defined by

$$\mathbf{x}'_\alpha = \frac{\partial \boldsymbol{\chi}_\alpha(\mathbf{X}_\alpha, t)}{\partial t} , \quad \mathbf{x}''_\alpha = \frac{\partial^2 \boldsymbol{\chi}_\alpha(\mathbf{X}_\alpha, t)}{\partial t^2} . \quad (2.10)$$

Using (2.5), the Eulerian description is gained for the velocity $\mathbf{v}_\alpha$ and the acceleration $\mathbf{a}_\alpha$:

$$\mathbf{v}_\alpha = \mathbf{x}'_\alpha = \mathbf{x}'_\alpha(\mathbf{x}, t) , \quad \mathbf{a}_\alpha = \mathbf{x}''_\alpha = \mathbf{x}''_\alpha(\mathbf{x}, t) . \quad (2.11)$$

As the individual constituents follow, in general, different motions, different material time derivatives must be formulated. This will be shown for an arbitrary scalar-value function $\Gamma(\mathbf{x}, t)$. Analogous material time derivatives of vector and tensor functions result. If $\Gamma(\mathbf{x}, t)$ is a differentiable function, then its material time derivative, following the motion of the constituent φ^α, is defined by

$$\Gamma'_\alpha = \frac{\partial \Gamma}{\partial t} + \mathrm{grad}\, \Gamma \cdot \mathbf{x}'_\alpha . \quad (2.12)$$

With $(2.10)_1$, the material velocity gradient of the constituent φ_α is obtained:

$$(\mathbf{F}_\alpha)'_\alpha = \mathrm{Grad}_\alpha \, x'_\alpha . \quad (2.13)$$

The spatial velocity gradient can be calculated from $(2.11)_1$ and results in

$$\mathbf{L}_\alpha = \operatorname{grad} \mathbf{x}'_\alpha = \operatorname{grad} \mathbf{v}_\alpha, \quad (2.14)$$

which is connected to the material velocity gradient and the deformation gradient by

$$\mathbf{L}_\alpha = (\mathbf{F}_\alpha)'_\alpha \mathbf{F}_\alpha^{-1}. \quad (2.15)$$

Usually, no distinction is made in the literature between $\mathbf{x}'_\alpha$ and $\mathbf{v}_\alpha$, as well as between $\mathbf{x}''_\alpha$ and $\mathbf{a}_\alpha$, because it is, in many cases, obvious in connection with the operator as to whether $\mathbf{x}'_\alpha(\mathbf{X}_\alpha, t)$ or $\mathbf{x}'_\alpha(\mathbf{x}, t)$, as well as $\mathbf{x}''_\alpha(\mathbf{X}_\alpha, t)$ or $\mathbf{x}''_\alpha(\mathbf{x}, t)$, is meant; see, for example, (2.13) and (2.14).

The additive decomposition of $\mathbf{L}_\alpha$ yields the symmetrical part $\mathbf{D}_\alpha$ of the spatial velocity gradient and the skew-symmetric spin tensor $\mathbf{W}_\alpha$

$$\mathbf{L}_\alpha = \mathbf{D}_\alpha + \mathbf{W}_\alpha \quad (2.16)$$

with

$$\mathbf{D}_\alpha = \frac{1}{2}(\mathbf{L}_\alpha + \mathbf{L}_\alpha^T), \quad \mathbf{W}_\alpha = \frac{1}{2}(\mathbf{L}_\alpha - \mathbf{L}_\alpha^T). \quad (2.17)$$

Since the local deformations $\mathbf{F}_\alpha$ contain, in general, parts of a rigid body motion, they are less suitable to act as measurements for the deformations in constitutive equations. For this reason, it is convenient to use the line-elements, in the form of the difference of the squares of the line-elements, in the actual and the reference placements for the measurement of the deformation, in order to avoid irrational operations and to take out the rigid body motions. For the evaluation of the squares of the line-elements, the transport mechanism $d\mathbf{x} = \mathbf{F}_\alpha d\mathbf{X}_\alpha$ gained from (2.7) will be used. After elementary calculations, the following relations are obtained:

$$d\mathbf{x} \cdot d\mathbf{x} - d\mathbf{X}_\alpha \cdot d\mathbf{X}_\alpha = d\mathbf{X}_\alpha \cdot 2\mathbf{E}_\alpha d\mathbf{X}_\alpha = d\mathbf{x} \cdot 2\mathbf{A}_\alpha d\mathbf{x}. \quad (2.18)$$

The introduced symmetric strain tensors $\mathbf{E}_\alpha$ and $\mathbf{A}_\alpha$ are known, respectively, as the Green strain tensor and the Almansi strain tensor. They depend on the deformation gradient $\mathbf{F}_\alpha$ in the following way:

$$\mathbf{E}_\alpha = \frac{1}{2}(\mathbf{C}_\alpha - \mathbf{I}), \quad \mathbf{A}_\alpha = \frac{1}{2}(\mathbf{I} - \mathbf{B}_\alpha^{-1}), \quad (2.19)$$

where

$$\mathbf{C}_\alpha = \mathbf{F}_\alpha^T \mathbf{F}_\alpha \quad \text{and} \quad \mathbf{B}_\alpha = \mathbf{F}_\alpha \mathbf{F}_\alpha^T \quad (2.20)$$

Considering (2.6), (2.15), and (2.16) the material time derivative of the Jacobian as well as of the right Cauchy-Green deformation tensor and the

Green strain tensor yields

$$(J_\alpha)'_\alpha = J_\alpha(\mathbf{D}_\alpha \cdot \mathbf{I}), \quad (\mathbf{C}_\alpha)'_\alpha = 2\,\mathbf{F}_\alpha^T \mathbf{D}_\alpha \mathbf{F}_\alpha, \quad (\mathbf{E}_\alpha)'_\alpha = \mathbf{F}_\alpha^T \mathbf{D}_\alpha \mathbf{F}_\alpha. \quad (2.21)$$

It is convenient for further considerations to express the Green strain tensor $\mathbf{E}_S$ by the displacement $\mathbf{u}_S$. With

$$\mathbf{x} = \mathbf{X}_S + \mathbf{u}_S \quad (2.22)$$

from (2.7) we obtain

$$\mathbf{F}_S = \mathbf{I} + \mathrm{Grad}_S \mathbf{u}_S = \mathbf{I} + \nabla \mathbf{u}_S, \quad (2.23)$$

and from (2.20)

$$\mathbf{C}_S = \mathbf{I} + \nabla \mathbf{u}_S + \nabla^T \mathbf{u}_S + \nabla^T \mathbf{u}_S \nabla \mathbf{u}_S. \quad (2.24)$$

In the case of small strains and small displacements and derivatives of the displacement vector (geometrically linear theory) we can neglect the quadratic term of the displacement gradients in (2.24) in comparison with the linear terms. Then, from (2.19) and (2.24)

$$\mathbf{E}_S = \frac{1}{2}(\nabla \mathbf{u}_S + \nabla^T \mathbf{u}_S) \quad (2.25)$$

is obtained. Moreover, within the geometrically linear theory the difference between the operators Grad_α and grad vanishes approximately and the spatial parameters $\mathbf{x}$ can be replaced by $\mathbf{X}_\alpha$.

In what follows, some considerations on the microscale are needed in order to describe the compressibility and incompressibility of the real materials. For this purpose, a macroscopic control space filled with a granular solid phase, and a gas without any physical properties, will be considered. The grains in the control space are represented by small balls. It is assumed that the grains are incompressible, i.e., a hydrostatic stress state in the grains produces no volume change of the grains. Although the grains are incompressible, contact forces acting on the grains cause a volume change of the control space; this results from the change of the pore structure and the volume fraction due to the change of the shapes of the individual grains. Therefore, the incompressibility condition cannot be expressed by the deformation gradient $\mathbf{F}_S$ of the partial solid constituent. Rather, the incompressibility condition must be formulated by physical quantities at the microscale. Moreover, statements on the compressibility and other real properties of the constituents must also be expressed by physical quantities at the microscale. In the case of describing compressibility and incompressibility, this means that a motion function at the microscale

$$\mathbf{x}_{SR(\text{micro})} = \mathbf{\chi}_{SR(\text{micro})}(\mathbf{X}_S + \boldsymbol{\xi}_{SR}, t) \quad (2.26)$$

must be introduced, where $\boldsymbol{\xi}_{SR}$ is the displacement vector. From (2.26), the deformation gradient $\mathbf{F}_{SR(\text{micro})}$ can be determined in a way similar to that in (2.7). Then, the incompressibility condition can be reformulated at the microscale:

$$\det \mathbf{F}_{SR(\text{micro})} = J_{SR(\text{micro})} = 1 \; . \quad (2.27)$$

The crucial point in this procedure is, however, the fact that the motion function $\boldsymbol{\chi}_{SR(\text{micro})}$ in (2.26) is completely unknown and cannot be determined by a balance principle within the framework of the mixture theory (macroscale). Therefore, it is advisable, in order to describe the phenomena of compressibility and incompressibility, to transfer the microscopic deformation behavior of the real solid phase to the macroscale. For this reason, the deformation tensor $\mathbf{F}_{SR}$ is introduced, which is understood to be part of the deformation gradient $\mathbf{F}_S$ and which is assumed to reflect the microscopic deformations of the real solid material at the macroscale. In general, the tensor $\mathbf{F}_{SR}$ is not integrable at the macroscale, i.e., the microscopic deformations $\mathbf{F}_{SR(\text{micro})}$ are represented by incompatible deformations at the macroscale. Since

$$\mathbf{F}_S \neq \mathbf{F}_{SR} \; , \quad (2.28)$$

it is necessary to choose a second tensor $\mathbf{F}_{SN}$, to transfer the relation (2.28) into an equation. The part $\mathbf{F}_{SN}$ of the deformation gradient $\mathbf{F}_S$, as well as $\mathbf{F}_{SR}$, is, in general, not integrable. On the contrary, the deformation tensor $\mathbf{F}_S$ is integrable; thus, the deformation gradient at the macroscale must be decomposed multiplicatively into $\mathbf{F}_{SR}$ and $\mathbf{F}_{SN}$.

In the following, some parallels to the theory of the elastic-plastic deformations of metals will be discussed. It is well-known that within the framework of a finite theory, a multiplicative decomposition of the deformation gradient into an elastic and a plastic part is widely used. The plastic part of the deformation at the macroscale results from dislocations at the microscale. These microscopic dislocations are also, in general, represented by incompatible strains at the macroscale. Therefore, the reason for a multiplicative decomposition of the deformation gradient at the macroscale is the same as in the porous media theory, namely to bring physical phenomena from the microscale to the macroscale.

The usefulness of the multiplicative decomposition of the deformation gradient in porous media theory will be revealed in the following paragraphs.

There are two possibilities of decomposing multiplicatively the deformation gradient $\mathbf{F}_\alpha$ of the constituent φ^α, of which only the following is suitable (see the extensive discussion of this problem and the consequences concerning the kinematics in Bluhm and de Boer, 1997):

$$\mathbf{F}_\alpha = \mathbf{F}_{\alpha N} \hat{\mathbf{F}}_{\alpha R} \,. \quad (2.29)$$

The part $\hat{\mathbf{F}}_{\alpha R}$ is interpreted as that part of $\mathbf{F}_\alpha$ which describes the deformation of the real material, whereas $\mathbf{F}_{\alpha N}$ describes the remaining part of the deformation of the control space, namely the change of the pores in size and shape. The parts $\mathbf{F}_{\alpha N}$ and $\hat{\mathbf{F}}_{\alpha R}$ are to be understood as local mappings of tangent (vector) spaces in each material point of the body. In the case of homogeneous deformations, the multiplicative decomposition (2.29) leads to an intermediate state $(\hat{\ldots})$.

The proof of the multiplicative decomposition (2.29) of $\mathbf{F}_\alpha$ into quantities describing properties of the microscale is still awaiting research.

In analogy to (2.15) through (2.17), material time derivatives of the deformation tensor $\hat{\mathbf{F}}_{\alpha R}$ can be introduced:

$$\hat{\mathbf{L}}_{\alpha R} = (\hat{\mathbf{F}}_{\alpha R})'_\alpha (\hat{\mathbf{F}}_{\alpha R})^{-1}\,, \quad \hat{\mathbf{L}}_{\alpha R} = \hat{\mathbf{D}}_{\alpha R} + \hat{\mathbf{W}}_{\alpha R}\,,$$

$$\hat{\mathbf{D}}_{\alpha R} = \frac{1}{2}(\hat{\mathbf{L}}_{\alpha R} + \hat{\mathbf{L}}_{\alpha R}^T)\,, \quad \hat{\mathbf{W}}_{\alpha R} = \frac{1}{2}(\hat{\mathbf{L}}_{\alpha R} - \hat{\mathbf{L}}_{\alpha R}^T)\,,$$

$$\hat{\mathbf{L}}_{\alpha R} = \frac{\partial (\hat{\mathbf{x}}_\alpha)'_\alpha}{\partial \mathbf{X}_\alpha} \frac{\partial \mathbf{X}_\alpha}{\partial \hat{\mathbf{x}}_\alpha} = \frac{\partial (\hat{\mathbf{x}}_\alpha)'_\alpha}{\partial \hat{\mathbf{x}}_\alpha} \quad \text{(homogeneous deformations)}\,. \quad (2.30)$$

The introduction of the material time derivative of $\mathbf{F}_{\alpha N}$, namely $\mathbf{L}_{\alpha N}$, is less useful because $\mathbf{L}_{\alpha N}$ is not a spatial velocity gradient (see Bluhm and de Boer, 1997).

Now, the volume fraction concept and the incompressibility condition under the consideration of the multiplicative decomposition of the deformation gradient $\mathbf{F}_\alpha$ of the constituent φ^α will be discussed. We proceed from (2.29) and split the two parts, $\mathbf{F}_{\alpha N}$ and $\hat{\mathbf{F}}_{\alpha R}$, of the deformation gradient into volume-preserving and spherical parts denoted by the symbols $(\tilde{\ldots})$ and $(\check{\ldots})$. Thus,

$$\mathbf{F}_{\alpha N} = \tilde{\mathbf{F}}_{\alpha N} \check{\mathbf{F}}_{\alpha N}\,, \quad \hat{\mathbf{F}}_{\alpha R} = \tilde{\mathbf{F}}_{\alpha R} \check{\mathbf{F}}_{\alpha R} \quad (2.31)$$

with

$$\tilde{\mathbf{F}}_{\alpha N} = (J_{\alpha N})^{1/3}\mathbf{I}, \quad J_{\alpha N} = \det \mathbf{F}_{\alpha N}, \quad \check{J}_{\alpha N} = \det \check{\mathbf{F}}_{\alpha N} = 1,$$

$$\tilde{\mathbf{F}}_{\alpha R} = (\hat{J}_{\alpha R})^{1/3}\mathbf{I}, \quad \hat{J}_{\alpha R} = \det \hat{\mathbf{F}}_{\alpha R}, \quad \check{J}_{\alpha R} = \det \check{\mathbf{F}}_{\alpha R} = 1. \quad (2.32)$$

With these quantities, kinematic expressions corresponding to those of $\tilde{\mathbf{F}}_\alpha$ and $\check{\mathbf{F}}_\alpha$ can be formulated. In particular, the following derivatives are valid:

$$(\hat{J}_{\alpha R})'_\alpha = \hat{J}_{\alpha R}(\hat{\mathbf{D}}_{\alpha R} \cdot \mathbf{I}), \quad (J_{\alpha N})'_\alpha = J_{\alpha N}(\mathbf{D}_{\alpha N} \cdot \mathbf{I}) \quad (2.33)$$

with

$$\mathbf{D}_{\alpha N} \cdot \mathbf{I} = \mathbf{D}_\alpha \cdot \mathbf{I} - \hat{\mathbf{D}}_{\alpha R} \cdot \mathbf{I}. \quad (2.34)$$

For further investigations, the volume elements in the reference and actual placements must be considered. In continuum mechanics, it is well-known that the following transport theorem concerning the volume elements is valid:

$$dv = J_\alpha dv_{0\alpha}, \quad (2.35)$$

where

$$dv_{0\alpha} = dv_{0\alpha}(\mathbf{X}_\alpha, t = t_0), \quad dv = dv(\mathbf{x}, t) \quad (2.36)$$

are the volume elements in the reference placement at the position $\mathbf{X}_\alpha$, denoted by the subscript index α, and in the actual placement at the position $\mathbf{x}$. In consideration of (2.29), (2.32), (2.35),

$$dv = J_{\alpha N}\hat{J}_{\alpha R}dv_{0\alpha} \quad (2.37)$$

is gained. By using (2.29), a differential volume $d\hat{v}_\alpha$, at a material point $\mathbf{X}_\alpha$ of a local intermediate placement in the tangent space, is related to the differential volume elements in the reference placement and the actual placements by (see de Boer, 1996, and Bluhm and de Boer, 1997)

$$d\hat{v}_\alpha = \hat{J}_{\alpha R}dv_{0\alpha}, \quad dv = J_{\alpha N}d\hat{v}_\alpha. \quad (2.38)$$

With the relations (2.37) and (2.38), it is easy to formulate various kinds of volume strains which will be important for the investigations in the following passages. In analogy to the volume strain of the partial material of the constituent φ^α,

$$e_\alpha = \frac{dv - dv_{0\alpha}}{dv_{0\alpha}} = \frac{dv}{dv_{0\alpha}} - 1 = J_\alpha - 1, \quad (2.39)$$

where (2.35) has been used, the volume strain of the real material of φ^α is defined as

$$e_{\alpha R} = \frac{dv^\alpha - dv_{0\alpha}^\alpha}{dv_{0\alpha}^\alpha} = \frac{n^\alpha dv - n_{0\alpha}^\alpha dv_{0\alpha}}{n_{0\alpha}^\alpha dv_{0\alpha}} = \frac{n^\alpha}{n_{0\alpha}^\alpha}\frac{dv}{dv_{0\alpha}} - 1$$

$$= \frac{n^\alpha}{n_{0\alpha}^\alpha} J_\alpha - 1 = \frac{n^\alpha}{n_{0\alpha}^\alpha} J_{\alpha N}\hat{J}_{\alpha R} - 1 \, , \quad (2.40)$$

where (2.1), (2.35), and (2.37) have been used. In consideration of the transport theorems (2.38), further real volume strains can be formulated:

$$\hat{e}_{\alpha R} = \frac{d\hat{v}_\alpha^\alpha - dv_{0\alpha}^\alpha}{dv_{0\alpha}^\alpha} = \frac{\hat{n}_\alpha^\alpha d\hat{v}_\alpha - n_{0\alpha}^\alpha dv_{0\alpha}}{n_{0\alpha}^\alpha dv_{0\alpha}} = \frac{\hat{n}_\alpha^\alpha}{n_{0\alpha}^\alpha}\frac{d\hat{v}_\alpha}{dv_{0\alpha}} - 1 = \frac{\hat{n}_\alpha^\alpha}{n_{0\alpha}^\alpha}\hat{J}_{\alpha R} - 1 \, ,$$

$$\tilde{e}_{\alpha R} = \frac{dv^\alpha - d\hat{v}_\alpha^\alpha}{d\hat{v}_\alpha^\alpha} = \frac{n^\alpha dv - \hat{n}_\alpha^\alpha d\hat{v}_\alpha}{\hat{n}_\alpha^\alpha d\hat{v}_\alpha} = \frac{n^\alpha}{\hat{n}_\alpha^\alpha}\frac{dv}{d\hat{v}_\alpha} - 1 = \frac{n^\alpha}{\hat{n}_\alpha^\alpha} J_{\alpha N} - 1 \, , \quad (2.41)$$

where the part $d\hat{v}_\alpha^\alpha$ of the differential volume element $d\hat{v}_\alpha$, in the local intermediate placement in the tangent space, is defined via the volume fraction $\hat{n}_\alpha^\alpha$ as

$$d\hat{v}_\alpha^\alpha = \hat{n}_\alpha^\alpha d\hat{v}_\alpha \, . \quad (2.42)$$

The determinants of $\hat{\mathbf{F}}_{\alpha R}$ and $\mathbf{F}_{\alpha N}$ can be expressed depending on the real volume strains $\hat{e}^{\alpha R}$ and $\tilde{e}^{\alpha R}$, namely

$$\hat{J}_{\alpha R} = \frac{n_{0\alpha}^\alpha}{\hat{n}_\alpha^\alpha}(\hat{e}_{\alpha R} + 1) \, , \quad J_{\alpha N} = \frac{\hat{n}_\alpha^\alpha}{n^\alpha}(\tilde{e}_{\alpha R} + 1) \, . \quad (2.43)$$

With (2.43), the real volume strain (2.40) can be reformulated as

$$e_{\alpha R} = \hat{e}_{\alpha R} + \tilde{e}_{\alpha R} + \hat{e}_{\alpha R}\tilde{e}_{\alpha R} \, . \quad (2.44)$$

The tensor $\hat{\mathbf{F}}_{\alpha R}$ is interpreted as that part of the deformation gradient which includes the whole deformation of the real material of the constituent φ^α. Thus, the determinant $J_{\alpha R}$ must reflect the volume strain of the real material of φ^α, i.e., the real volume strain is the difference between the part $d\hat{v}_\alpha^\alpha = \hat{n}_\alpha^\alpha d\hat{v}_\alpha$ of the differential volume $d\hat{v}_\alpha$ in the local intermediate placement in the vector space, and the part $dv_{0\alpha}^\alpha = n_{0\alpha}^\alpha dv_{0S}$ of the volume element dv_{0S} in the reference placement at the position $\mathbf{X}_S$. Therefore, the following relations concerning the real volume strains hold:

$$e_{\alpha R} = \hat{e}_{\alpha R} \, , \quad \tilde{e}_{\alpha R} = 0 \, . \quad (2.45)$$

Furthermore, the transport theorem $(2.38)_1$ excludes the change of the volume fraction by mapping $dv_{0\alpha}$ from the reference to the local intermediate placement, i.e.,

$$\hat{n}_\alpha^\alpha = n_{0\alpha}^\alpha . \quad (2.46)$$

With (2.45) and (2.46), the determinants $\hat{J}_{\alpha R}$ and $J_{\alpha N}$, see (2.43), read as follows:

$$\hat{J}_{\alpha R} = \hat{e}_{\alpha R} + 1 = e_{\alpha R} + 1 , \quad J_{\alpha N} = \frac{n_{0\alpha}^\alpha}{n^\alpha} . \quad (2.47)$$

In the case of incompressibility,

$$e_{\alpha R} = 0 , \quad \hat{J}_{\alpha R} = 1 , \quad \text{and} \quad \hat{\mathbf{D}}_{\alpha R} \cdot \mathbf{I} = 0 \quad (2.48)$$

are valid, whereby $(2.48)_3$ is the rate formulation of the incompressibility condition (see Bluhm and de Boer, 1997) and where $\hat{\mathbf{D}}_{\alpha R}$ can be interpreted as the Lie derivative of the strain tensor $\mathbf{E}_{\alpha R} = \frac{1}{2}(\mathbf{F}_{\alpha R}^T \mathbf{F}_{\alpha R} - \mathbf{I})$.

With the help of the multiplicative decomposition of the deformation gradient $\mathbf{F}_\alpha$ (2.29), it can be shown that the statement of Mills (1967) and Bowen (1980) concerning the incompressibility of the real material, which was described by setting the real densities of the constituents constant, is only identical with the kinematic constraints (2.48) in the case of a thermodynamic process without any mass exchange (see Bluhm and de Boer, 1997).

As there is no difference in the volume change of the real material in the actual and intermediate placements, we will omit the superscript sign $(\hat{\ldots})$ in the following sections.

Moreover, it is – in the geometrically linear theory – sometimes convenient to decompose the partial volumetric strain e_S into e_{SR}, e_{SN}

$$e_S = e_{SR} + e_{SN}. \quad (2.49)$$

Now, we will leave the development of kinematics and will turn to the formulation of balance principles.

2.4 Balance principles

In mixture theory and porous media theory, balance principles – balance of mass, balance of momentum, and moment of momentum, as well as balance of energy – must be established for each constituent φ^α in consideration of all interaction and external agencies, as had already been called for by von Terzaghi (1925). This means that all quantities resulting from long- and short-range effects which influence the individual constituents, as well as the interaction effects between the constituents, have to be considered in the balance principles.

In the time following, balance principles have been discussed at length by Truesdell (1957a, b, 1969, 1984), Truesdell and Toupin (1960), Kelly (1964), Eringen and Ingram (1965), Green and Naghdi (1965), Müller (1968), Bowen and Wiese (1969), and de Boer and Ehlers (1986).

The interaction effects (supply terms) have to be in sum equal to zero. These conditions are at least founded by the fact that – in the case of a common velocity **v** and acceleration **a**, a common external acceleration **b**, as well as a common internal energy ε, and external heat supply r – the sum of balance principle must formally become the corresponding balance principles of a one-component body.

The following balance principles will be formulated in global and local forms for the individual constituents in consideration of all interaction effects and, finally, some conclusions in view of the comparison with the balance principle of the mixture body will be drawn.

2.5 Balance of Mass

As has already been mentioned, two possibilities exist concerning the formulation of the balance principle of mass. On the one hand, this equation can be given for the bulk mixture body; on the other hand, the mass balance principles can be formulated for each individual constituent in such a way that the superposition of the mass balance principles for the individual constituents turn, for special cases, into the balance principle of the mixture body as a one-component body.

The balance of mass for the individual constituents φ^α requires that the rate of the mass M^α equals a mass term $\int_{B_\alpha} \hat{\rho}^\alpha dv$ caused by the other constituents, where $\hat{\rho}^\alpha$ is the mass supply per volume element:

$$(M^\alpha)'_\alpha = \left(\int_{B_\alpha} \rho^\alpha dv \right)'_\alpha = \int_{B_\alpha} \hat{\rho}^\alpha dv \ . \quad (2.50)$$

The integration in (2.50) covers the domain B_α of each individual constituent. With the help of the transport theorem

$$(dv)'_\alpha = \operatorname{div} \mathbf{v}_\alpha dv \quad (2.51)$$

from (2.50), the local statement

$$(\rho^\alpha)'_\alpha + \rho^\alpha \operatorname{div} \mathbf{v}_\alpha = \hat{\rho}^\alpha \quad (2.52)$$

or

$$\frac{\partial \rho^\alpha}{\partial t} + \operatorname{div}(\rho^\alpha \mathbf{v}_\alpha) = \hat{\rho}^\alpha \quad (2.53)$$

is derived. Assuming a common velocity $\mathbf{v}$ for all phases φ^α, the summation of (2.53) over all κ constituents φ^α leads to

$$\sum_{\alpha=1}^{\kappa} \left[\frac{\partial \rho^\alpha}{\partial t} + \operatorname{div}(\rho^\alpha \mathbf{v}) \right] = \sum_{\alpha=1}^{\kappa} \hat{\rho}^\alpha . \quad (2.54)$$

With the statement that the sum of the densities of the individual constituents is equal to the density ρ of the mixture body,

$$\rho = \sum_{\alpha=1}^{\kappa} \rho^\alpha , \quad (2.55)$$

one obtains

$$\frac{\partial \rho}{\partial t} + \operatorname{div}(\rho \mathbf{v}) = \sum_{\alpha=1}^{\kappa} \hat{\rho}^\alpha . \quad (2.56)$$

For deriving Eqn. (2.56), we have made use of the fact that the summation and the derivative are interchangeable. Eqn. (2.56) only results in the form valid for the mixture body if the constraint

$$\sum_{\alpha=1}^{\kappa} \hat{\rho}^\alpha = 0 \quad (2.57)$$

is introduced, i.e., if the sum of the local mass supplies of all κ constituents φ^α is equal to zero.

If all mass exchange is excluded, the relation (2.52) can be integrated and it follows that:

$$\rho^\alpha = \rho^\alpha_{0\alpha} (\det \mathbf{F}_\alpha)^{-1} \quad \text{with} \quad \rho^\alpha_{0\alpha} = \rho^\alpha_{0\alpha}(\mathbf{X}_\alpha, t = t_0) . \quad (2.58)$$

The quantity $\rho^\alpha_{0\alpha}$ denotes the partial density of the constituent φ^α (superscript index) in the reference placement at the position $\mathbf{X}_\alpha$.

Substituting $\mathbf{F}_L$ by the displacement vector $\mathbf{u}_L$ according to (2.23) we have

$$\rho^L = \rho^L_{0L} \det(\mathbf{I} + \nabla \mathbf{u}_L)^{-1} = (1 + \nabla \mathbf{u}_L \cdot \mathbf{I})^{-1} \rho^L_{0L}$$
$$= (1 - \nabla \mathbf{u}_L \cdot \mathbf{I}) \rho^L_{0L}, \quad (2.59)$$

within the geometrically linear theory. Taking the gradient of ρ^L we arrive at

$$\operatorname{grad} \rho^L = -\rho^L_{0L} (\nabla \nabla \mathbf{u}_L)^T \overset{13}{\mathbf{I}}, \quad (2.60)$$

which is needed in the constitutive equation for the capillary forces.

2.6   Balance of Momentum and Moment of Momentum

In this section, the consequences of applying the axioms of the *balance of momentum* to each individual constituent will first be discussed.

The balance principle of momentum states that the material time derivative of momentum is equal to the sum of external forces. Thus,

$$(\mathbf{l}^\alpha)'_\alpha = \mathbf{k}^\alpha. \quad (2.61)$$

Herein, the momentum $\mathbf{l}^\alpha$ for the constituent φ^α is defined by

$$\mathbf{l}^\alpha = \int_{B_\alpha} \rho^\alpha \mathbf{v}_\alpha dv. \quad (2.62)$$

The external forces $\mathbf{k}^\alpha$ are given by the sum of the forces $\mathbf{f}^\alpha$, which are caused by long- and short-range effects and are acting on the constituents as volume forces $\rho^\alpha \mathbf{b}^\alpha$ and surface forces $\mathbf{t}^\alpha$, as well as the interaction forces $\hat{\mathbf{p}}^\alpha$ which belong to the volume forces. The resulting force vector $\mathbf{k}^\alpha$ is thus given by

$$\mathbf{k}^\alpha = \mathbf{f}^\alpha + \int_{B_\alpha} \hat{\mathbf{p}}^\alpha dv, \quad (2.63)$$

where $\mathbf{f}^\alpha$ is composed of

$$\mathbf{f}^\alpha = \int_{B_\alpha} \rho^\alpha \mathbf{b}^\alpha dv + \int_{\partial B_\alpha} \mathbf{t}^\alpha da. \quad (2.64)$$

Now, all the terms which are contained in the balance principle of momentum have been listed. With Cauchy's theorem,

$$\mathbf{t}^\alpha = \mathbf{T}^\alpha \mathbf{n}, \quad (2.65)$$

where $\mathbf{T}^\alpha$ is Cauchy's stress tensor of the constituent φ^α and $\mathbf{n}$ the unit normal at the surface of the individual constituent body. With the divergence theorem, as well as with either mass balance principle, (2.52) or (2.53), Cauchy's first equation of motion (balance of momentum) for φ^α is obtained from (2.61):

$$\operatorname{div} \mathbf{T}^\alpha + \rho^\alpha \mathbf{b}^\alpha + \hat{\mathbf{p}}^\alpha = \rho^\alpha \mathbf{a}_\alpha + \hat{\rho}^\alpha \mathbf{v}_\alpha . \quad (2.66)$$

In this equation, the expression $\hat{\rho}^\alpha \mathbf{v}_\alpha$ represents the exchange of linear momentum through the density supply $\hat{\rho}^\alpha$.

The balance of momentum for the mixture body can be gained by superimposition of the momenta of all κ constituents φ^α, assuming a common velocity $\mathbf{v}$ and acceleration $\mathbf{a}$, as well as a common external acceleration $\mathbf{b}$ for all constituents φ^α:

$$\sum_{\alpha=1}^{\kappa} \left(\operatorname{div} \mathbf{T}^\alpha + \rho^\alpha \mathbf{b} + \hat{\mathbf{p}}^\alpha \right) = \sum_{\alpha=1}^{\kappa} \left(\rho^\alpha \mathbf{a} + \hat{\rho}^\alpha \mathbf{v} \right) . \quad (2.67)$$

Introducing the requirements

$$\mathbf{T} = \sum_{\alpha=1}^{\kappa} \mathbf{T}^\alpha, \quad \rho \mathbf{b} = \sum_{\alpha=1}^{\kappa} \rho^\alpha \mathbf{b},$$

$$\dot{\mathbf{l}} = \sum_{\alpha=1}^{\kappa} [\rho^\alpha \mathbf{a} + \hat{\rho}^\alpha \mathbf{v}] = \rho \mathbf{a}, \quad \sum_{\alpha=1}^{\kappa} \hat{\mathbf{p}}^\alpha = \mathbf{0} , \quad (2.68)$$

where $\mathbf{T}, \rho \mathbf{b}$, and $\dot{\mathbf{l}}$ are, respectively, Cauchy's stress tensor, the volume force, and the time rate of the momentum of the mixture body, the balance principle of the mixture body is gained:

$$\operatorname{div} \mathbf{T} + \rho \mathbf{b} = \dot{\mathbf{l}} . \quad (2.69)$$

The material time derivative $(\dot{\ldots})$ has to be formed with the velocity $\mathbf{v}$. The requirement of $(2.68)_4$ is a constraint for the momentum supplies. Note that the second summation in $(2.68)_3$ vanishes due to (2.57).

The *balance of moment of momentum* for non-polar materials states that the material time derivative of the moment of momentum is equal to the moments of all external forces, where the moments are referred to a fixed point 0:

$$\left(\mathbf{h}_{(0)}^\alpha \right)'_\alpha = \mathbf{m}_{(0)}^\alpha . \quad (2.70)$$

We do not consider local moment of momentum supply vectors $\hat{\mathbf{m}}^\alpha$ or the corresponding tensors $\hat{\mathbf{M}}^\alpha$. For the formulation of the moment of momentum balance principle for polar materials, the reader is referred to Ehlers and Volk (1999).

The moment of momentum $\mathbf{h}_{(0)}^\alpha$ for the constituent φ^α is given in consideration of (2.63) by:

$$\mathbf{h}_{(0)}^\alpha = \int_{B_\alpha} \mathbf{x} \times \rho^\alpha \mathbf{v}_\alpha \, dv \,. \quad (2.71)$$

The moment of the external forces $\mathbf{m}_{(0)}^\alpha$ can be calculated considering (2.63) and (2.64) from:

$$\mathbf{m}_{(0)}^\alpha = \int_{B_\alpha} \mathbf{x} \times (\rho^\alpha \mathbf{b}^\alpha + \hat{\mathbf{p}}^\alpha) \, dv + \int_{\partial B_\alpha} \mathbf{x} \times \mathbf{t}^\alpha \, da \,. \quad (2.72)$$

From (2.70), considering (2.71) and (2.72) as well as the mass balance (2.52), the material time derivative of the moment of momentum leads to

$$\left(\mathbf{h}_{(0)}^\alpha\right)'_\alpha = \int_{B_\alpha} \mathbf{x} \times (\rho^\alpha \mathbf{a}_\alpha + \hat{\rho}^\alpha \mathbf{v}_\alpha) \, dv \,. \quad (2.73)$$

Moreover, the evaluation of the expression for the moment $\mathbf{m}_{(0)}^\alpha$, considering the balance principle of momentum (2.66), yields:

$$\mathbf{m}_{(0)}^\alpha = \int_{B_\alpha} \mathbf{x} \times (\rho^\alpha \mathbf{a}_\alpha + \hat{\rho}^\alpha \mathbf{v}_\alpha) \, dv + \int_{B_\alpha} \mathbf{I} \times \mathbf{T}^\alpha \, dv \quad (2.74)$$

from which, with (2.70) and (2.73), the local statement

$$\mathbf{I} \times \mathbf{T}^\alpha = \mathbf{0} \quad (2.75)$$

is obtained.

The above statement is fulfilled, if

$$\mathbf{T}^\alpha = (\mathbf{T}^\alpha)^T, \quad (2.76)$$

i.e., if Cauchy's stress tensor is symmetric.

The requirement that the mixture, as the sum of all κ constituents, should behave as a one-component material contains the condition:

$$\sum_{\alpha=1}^{\kappa} \mathbf{T}^\alpha = \sum_{\alpha=1}^{\kappa} (\mathbf{T}^\alpha)^T , \quad \mathbf{T} = \mathbf{T}^T , \quad (2.77)$$

whereby $(2.68)_1$ has been considered. The result of the balance of moment of momentum is the evaluation of the statement that also the stress tensor of the mixture body is symmetric.

3 Constitutive Equations

3.1 Constitutive Theory

As has already been mentioned, in order to close the system of field equations, it is necessary to introduce constitutive equations. These equations connect certain mechanical or thermodynamic quantities via material-dependent constants which are determined by test observations. Thus, it is ensured that the constitutive relations introduced are able to describe the test results.

In the past, a great number of constitutive equations for empty and saturated porous media have been derived. However, many of them are very complicated due to the use of inadequate mechanical or thermodynamic concepts. Without a doubt, those constitutive equations may closely describe the stress-strain (rate) relations of the special mechanical behavior of materials. However, in many cases this can only be achieved by introducing many parameters and neglecting requirements due to mechanical and thermodynamic "principles". Those constitutive equations are meaningless in view of the calculation of general boundary- and initial-value problems within the framework of geometrically-linear and non-linear theories. The goal should be to formulate relatively simple constitutive equations. This statement becomes even more evident in the field of saturated porous media. In this field, not only the thermomechanical behavior of the skeleton and the content of the pores have to be described, but also various interaction phenomena. Therefore, the idea of formulating relatively simple constitutive equations is of particular relevance. In order to derive consistent constitutive equations that are relatively simple, some strong assumptions have to be introduced at times. For example, for many problems the compressibility of the solid matrix material can be neglected in comparison with the compressibility of the matrix. Thus, the mathematical model reduces to an incompressible model. Another example is the elastic model, where also some simplifying assumptions have to be introduced in order to predict the essential properties of the porous solids under study by using relatively simple constitutive relations.

However, it is not sufficient to only fulfill requirements due to test observations; rather more general "principles", which were developed in continuum mechanics in the 1950s and 1960s, should be fulfilled; these are: *Determinism, local action, material objectivity,* and *dissipation*. Some of the above-stated "principles", however, should not be understood as axioms, but should rather be considered as convenient work hypotheses, because the real mechanical or thermodynamic behavior of solids and fluids is, in most cases, too complex to be described by relatively simple "principles".

The most important of the above-stated "principles" are the material objectivity and dissipation principles. The material objectivity principle states that the constitutive equations have to be formulated in such a way that they are not influenced by superimposed rigid body motions. The dissipation principle results from the second law of thermodynamics. Both principles have a big impact on the development of consistent constitutive equations.

Guided by these principles, constitutive equations for compressible and incompressible elastic porous solids, filled with compressible gas and incompressible liquid will be stated in the following sections, where it is assumed that the incompressible liquid acts as a carrier for the gas in the shape of bubbles. The counterpart, namely gas as a carrier for the liquid (droplets), is also possible, but however needs certain conversions. The aforementioned individual "principles" will be fulfilled – though not explicitly discussed – here.

3.2 The Closure Problem and the Saturation Constraint

Mixture theory – the basis of porous media theory – is closed, i.e., the number of unknown fields is equal to the sum of the balance principles and the constitutive relations. This can easily be proven. However, by the introduction of the volume fractions n^S, n^L and n^G for the real constituents φ^S, φ^L and φ^G in the porous media theory (in order to obtain homogenized (smeared) continua which can be treated by continuum mechanical methods), a problem arises in which two field equations are missing. This causes a considerable difference between the porous media theory and the mixture theory, as well as the continuum mechanics of one-component materials. Also, other existing theories in continuum mechanics are closed and every new condition leads to an equation in excess. This condition must be provided with a Lagrange multiplier for the evaluation process of the entropy inequality. If the equation in excess is a constraint of motion, then the Lagrange multiplier will become an unknown reaction force.

In the porous media theory, on the other hand, one has to look for additional equations in order to close the fields. It is however difficult to gain additional fields since the volume fractions contain quantities of the microscale for which balance or constitutive equations are not contained in the macroscopic mixture theory. Therefore, much effort has been made to overcome this crucial problem. This effort starts by introducing an additional balance principle to the formulation of an evolution equation for the volume fraction. This procedure solves – from the mathematical point of view – the closure problem. However – from the physical point of view – this method is completely insufficient, because one must be aware of an important constraint, namely the saturation condition (2.3). This constraint restricts, in the rate formulation, the rates of the volumetric changes and must, therefore, be considered in the evaluation of the entropy inequality. By differentiating the saturation condition (2.3) with respect to the solid phase (the same result can be obtained by differentiating with respect to the fluid phase), we have

$$(n^S)'_S + (n^L)'_S + (n^G)'_S = 0 \quad (3.1)$$

or

$$-(n^S)'_S - (n^L)'_L - (n^G)'_G + $$
$$+ \operatorname{grad} n^L \cdot (\mathbf{v}_L - \mathbf{v}_S) + \operatorname{grad} n^G \cdot (\mathbf{v}_G - \mathbf{v}_S) = 0. \quad (3.2)$$

Considering (2.2), we obtain from (3.2):

$$-n^S \frac{(\rho^S)'_S}{\rho^S} + n^S \frac{(\rho^{SR})'_S}{\rho^{SR}} - n^L \frac{(\rho^L)'_L}{\rho^L} + n^L \frac{(\rho^{LR})'_L}{\rho^{LR}} -$$
$$- n^G \frac{(\rho^G)'_G}{\rho^G} + n^G \frac{(\rho^{GR})'_G}{\rho^{GR}} + \operatorname{grad} n^L \cdot (\mathbf{v}_L - \mathbf{v}_S) +$$
$$+ \operatorname{grad} n^G \cdot (\mathbf{v}_G - \mathbf{v}_S) = 0 \quad (3.3)$$

or considering the mass balance principles (2.52), in connection with (2.2), (2.21)$_1$, (2.31), (2.32), (2.33), and (2.47), as well as neglecting the mass supplies,

$$n^S (\mathbf{D}_{SN} \cdot \mathbf{I}) + n^L (\mathbf{D}_{LN} \cdot \mathbf{I}) + n^G (\mathbf{D}_{GN} \cdot \mathbf{I}) +$$
$$+ \operatorname{grad} n^L \cdot (\mathbf{v}_L - \mathbf{v}_S) + \operatorname{grad} n^G \cdot (\mathbf{v}_G - \mathbf{v}_S) = 0 \quad (3.4)$$

or

$$n^S(\mathbf{D}_S \cdot \mathbf{I}) - n^S(\mathbf{D}_{SR} \cdot \mathbf{I}) + n^L(\mathbf{D}_L \cdot \mathbf{I}) -$$
$$-n^L(\mathbf{D}_{LR} \cdot \mathbf{I}) + n^G(\mathbf{D}_G \cdot \mathbf{I}) - n^G(\mathbf{D}_{GR} \cdot \mathbf{I}) +$$
$$+ \operatorname{grad} n^L \cdot (\mathbf{v}_L - \mathbf{v}_S) + \operatorname{grad} n^G \cdot (\mathbf{v}_G - \mathbf{v}_S) = 0 \quad (3.5)$$

is obtained. In (3.4) and (3.5), use is made of the relations

$$\frac{(n^\alpha)'_\alpha}{n^\alpha} + \mathbf{D}_{\alpha N} \cdot \mathbf{I} = 0 \, , \quad \mathbf{D}_{\alpha N} \cdot \mathbf{I} = \mathbf{D}_\alpha \cdot \mathbf{I} - (\mathbf{D}_{\alpha R} \cdot \mathbf{I}) \quad (3.6)$$

and

$$\frac{(\rho^{\alpha R})'_\alpha}{\rho^{\alpha R}} + \mathbf{D}_{\alpha R} \cdot \mathbf{I} = 0 \, , \quad (3.7)$$

which results from the balance principles of mass (2.52) or (2.53), excluding all mass exchanges, considering (2.2) and (2.33) (see Bluhm, 1997).

The relation (3.5) clearly reveals that the rates of the volumetric strains of the partial bodies $\mathbf{D}_\alpha \cdot \mathbf{I}$ and of the real compressible materials $\mathbf{D}_{\alpha R} \cdot \mathbf{I}$ are dependent. Please note in passing that, we have omitted the signs of $\mathbf{D}_{SR}$ and $\mathbf{D}_{FR}$ which characterize the intermediate state at this point and in the following section for the sake of simplicity.

The problem to be solved depends as to whether the constraint in the forms (3.2), (3.3), or (3.4), (3.5) should be used. If all mass exchanges are neglected, then the constraints in the forms (3.4) and (3.5) are convenient. If, however, mass exchange occurs, then the forms (3.2) and (3.3) have to be used in the evaluation of the entropy inequality.

As has already been mentioned, the saturation constraints (3.2) and (3.3), or (3.4) and (3.5) have to be considered in the evaluation of the entropy inequality because the rates of either the densities or of the volumetric strains of the solid and the fluid as well as the gas phases are dependent. In order to obtain a stress-power-like expression, the constraints (3.2) and (3.3), or (3.4) and (3.5), which contain the rates of the volumetric strains, will be multiplied by a hydrostatic interface pressure λ. It is true that the saturation condition (2.3) is an equation to further reduce the number of unknown volume fractions, but the grade of indetermination does not change by the introduction of the interface pressure λ. Therefore, it will be postulated that two constitutive equations for λ must be introduced (or two constitutive relations for the hydrostatic pressure in the solid material) which contain properties of both the constituents of the partial solid and of partial fluid phases, in order to achieve closure. This is a reasonable demand from the mechanical point of view, because the interface pressure acts in the solid as well as the fluid and gas phases. It will be seen that the introduced requirement leads to excellent physical results.

If the materials of the three individual constituents behave as incompressible phases, then additional constraints have to be considered in the evaluation of the entropy inequality.

The main result of the constitutive theory are the following constitutive relations for the stress of the compressible elastic solid phase $\mathbf{T}^S$ (partial stress) and p^{SR} (real hydrostatic stress of solid material) and the interaction forces $\hat{\mathbf{p}}^S, \hat{\mathbf{p}}^L, \hat{\mathbf{p}}^G$, for further information see de Boer (2000), de Boer and Didwania (2002):

$$\mathbf{T}^S = -n^S p \, \mathbf{I} + \mathbf{T}_E^S, \quad (3.8)$$

where p is the liquid or gas pressure, and the effective stresses $\mathbf{T}_E^S$ are restricted by the constitutive equation

$$\mathbf{T}_E^S = -\rho^S \frac{\partial \psi^S}{\partial \mathbf{E}_S}, \quad (3.9)$$

where ψ^S is the free Helmholtz energy function. The additive decomposion of the total stress $\mathbf{T}^S$ in a part of the liquid or gas pressure p and the effective stress, which depends on the motion of the partial solid, is a consequence of the saturation constraint (see de Boer, 2000).

With an appropriate ansatz for the free Helmholtz energy function and the additive decomposition of the strain tensor $\mathbf{E}_S$ in a deviatoric part $\mathbf{E}_S^D$ and a volumetric part $(\mathbf{E}_S \cdot \mathbf{I})\mathbf{I}$, we obtain

$$\mathbf{T}_E^S = 2\mu^S \mathbf{E}_S^D + K^S (\mathbf{E}_S \cdot \mathbf{I})\mathbf{I}, \quad (3.10)$$

where the partial compression modulus K^S is connected with the Lamé constants μ^S, λ^S through

$$K^S = \frac{2}{3}\mu^S + \lambda^S. \quad (3.11)$$

It is desirable to express the partial material constant K^S through quantities of the real solid material and of the empty porous solid. On the one hand these material parameters can in many cases be determined experimentally in a simple way, on the other hand this procedure points the immediate consideration of the incompressibility of the real material. This is however not possible for the Lamé constant μ^S because this would require the consideration of the shearing of the real material. Because there is no balance principle or constitutive relation for the real shearing a split of $\mathbf{E}_S^D$ is not meaningful.

However, the situation concerning the compression modulus is completely different because there exists a constitutive equation for the volumetric strain of the real solid material. This important feature of the constitutive theory will be revealed in the following paragraphs.

The constitutive equation (3.10) is the result of the exploitation of the entropy inequality if the free Helmholtz energy function is assumed to be dependent on the strain tensor $\mathbf{E}_S$. The volumetric strain $e_S = \mathbf{E}_S \cdot \mathbf{I}$ can be decomposed according to (2.49) into e_{SR} and e_{SN}. These volumetric strains are completely independent and it is advisable to require that the free Helmholtz energy function depends on e_{SR} and e_{SN} instead of e_S because it can be expected that one gets more information in the new constitutive equation gaining from the evaluation of the entropy inequality. From the entropy inequality we obtain with the new parameters e_{SR} and e_{SN} the following two relations for the hydrostatic stress and strain states (see de Boer, 2000):

$$p^S = \frac{1}{3}(\mathbf{T}^S \cdot \mathbf{I}) = -n^S p + n^S p_E^{SN}, \quad (3.12)$$

where p_E^{SN} depends on e_{SN} and

$$p^S = n^S p_E^{SR}, \quad (3.13)$$

where p_E^{SR} is dependent on e_{SR}.

Within the geometrical linear theory the following constitutive equations are proposed

$$p_E^{SN} = K^{SN} e_{SN} \quad (3.14)$$

and

$$p_E^{SR} = K^{SR} e_{SR} \quad (3.15)$$

The compression moduli K^{SN} and K^{SR}, representing the compression modulus of the empty porous solid and the compression modulus of the real solid material which can be easily experimentally determined (see Biot and Willis 1957). With (3.14) and (3.15) the constitutive equations (3.12) and (3.13) can be rewritten:

$$p^S = -n^S p + n^S K^{SN} e_{SN}, \quad (3.16)$$

$$p^S = n^S K^{SR} e_{SR}. \quad (3.17)$$

In order to bring the constitutive equations (3.16) and (3.17) into the form (3.10), where the stress and the strain tensors are partial quantities, some manipulations must be performed.

At first the real volumetric strain e_{SR} will be replaced with (2.49)

$$e_{SR} = e_S - e_{SN}. \quad (3.18)$$

Then, from (3.16) and (3.17) combined with (3.18) we have

$$n^S K^{SR} e_S - n^S K^{SR} e_{SN} = -n^S p + n^S K^{SN} e_{SN} \quad (3.19)$$

or

$$e_{SN} = \frac{1}{K^{SR} + K^{SN}} (p + K^{SR} e_S). \quad (3.20)$$

With (3.20) we can determine the partial hydrostatic pressure p^S (3.16)

$$p^S = -n^S p \left(1 - \frac{K^{SN}}{K^{SR} + K^{SN}}\right) + n^S \frac{K^{SN} K^{SR}}{K^{SR} + K^{SN}} e_S \quad (3.21)$$

or

$$p^S = -n^S \frac{1}{1 + \frac{K^{SN}}{K^{SR}}} p + n^S \frac{K^{SN}}{1 + \frac{K^{SN}}{K^{SR}}} e_S \quad (3.22)$$

In the special case, if the solid material is incompressible – that means $K^{SR} \to \infty$ – the relations (3.21) or (3.22) turn to

$$p^S = -n^S p + n^S K^{SN} e_S. \quad (3.23)$$

Finally, we obtain from (3.8), (3.10), (3.12) and (3.21) with

$$\mathbf{T}^S = \mathbf{T}^{SD} + \frac{1}{3}(\mathbf{T}^S \cdot \mathbf{I}) = \mathbf{T}^{SD} + p^S\mathbf{I}, \quad (3.24)$$

$$\mathbf{T}^S = -n^S p\left(1 - \frac{K^{SN}}{K^{SR} + K^{SN}}\right)\mathbf{I} + 2\mu^S \mathbf{E}_S^D$$
$$+ n^S \frac{K^{SN}K^{SR}}{K^{SR} + K^{SN}} e_S. \quad (3.25)$$

or

$$\mathbf{T}^S = -n^S p\left(1 - B^S\right)\mathbf{I} + 2\mu^S \mathbf{E}_S^D + \hat{K}^S e_S, \quad (3.26)$$

where we have introduced the abbreviations

$$B^S = \frac{K^{SN}}{K^{SR} + K^{SN}}, \quad (3.27)$$

$$\hat{K}^S = n^S K^{SR} B^S. \quad (3.28)$$

For further considerations, namely for the development of the fundamental field equations in elastodynamics a slight reformulation of (3.26) is necessary in order to be able to compare the constitutive equation for the elastic behavior of the solid skeleton and the field equations with those of the classical linear elasticity theory. For this reason we replace $\mathbf{E}_S^D$ with $\mathbf{E}_S$ in (3.26). This procedure leads to

$$\mathbf{T}^S = -n^S p\left(1 - B^S\right)\mathbf{I} + 2\mu^S \mathbf{E}_S + \hat{\lambda}^S e_S \mathbf{I}, \quad (3.29)$$

with

$$\hat{\lambda}^S = \hat{K}^S - \frac{2}{3}\mu^S. \quad (3.30)$$

It should be mentioned that the quantity B^S (3.27) in the first term of the right-hand of (3.26) and (3.29) is similar to the Biot factor (see Biot and Willes, 1957) with the difference being that the denominator in the Biot factor contains only the compression modulus of the real material K^{SR} whereas in

(3.27) the denominator consists of the sum of the compression modulus of the real materials K^{SR} and the compression modulus of the solid skeleton K^{SN}. Only in the case where K^{SN} is very small in comparison with K^{SR}, both results are approximately equal.

Moreover, for the inviscid liquid and gas phases thermodynamic investigations reveal the following constitutive equations, if the liquid phase is considered as the carrier of the gas bubbles.

$$\mathbf{T}^L = -n^L p\, \mathbf{I} \quad (3.31)$$

and

$$\mathbf{T}^G = -n^G p\, \mathbf{I} + \mathbf{T}^G_E, \quad (3.32)$$

where

$$\mathbf{T}^G_E = p^G_E \mathbf{I} \quad (3.33)$$

with

$$\mathbf{p}^G_E = \rho^G \frac{\partial \psi^G}{\partial n^G} \quad (3.34)$$

is the effective stress for the gas phase.

With appropriate constitutive ansätze for (3.34), p^G_E can be expressed by the displacement vector $\mathbf{u}_G$. However, there is little experience with these constitutive equations until now. Thus, we maintain p^G_E in the fundamental equations.

In order to be able to formulate the fundamental field equations in poroelasticity in the next section also the constitutive equations for the interaction force $\hat{\mathbf{p}}^S$ must be addressed. This has been done by de Boer and Didwania (2002) for a capillary-porous solid field with liquid and gas

$$\hat{\mathbf{p}}^S = p\, \text{grad}\, n^S - \hat{t}^{SL}\, \text{grad}\, \rho^L + S_L(\mathbf{v}_L - \mathbf{v}_S) - S_G(\mathbf{v}_G - \mathbf{v}_S), \quad (3.35)$$

$$\hat{\mathbf{p}}^L = p\, \text{grad}\, n^L + (\hat{t}^{SL} + \hat{t}^{LG})\, \text{grad}\, \rho^L - S_L(\mathbf{v}_L - \mathbf{v}_S), \quad (3.36)$$

$$\hat{\mathbf{p}}^G = p \text{ grad } n^G - \hat{t}^{LG} \text{grad} \rho^L + S_G(\mathbf{v}_G - \mathbf{v}_S). \quad (3.37)$$

In (3.35) to (3.37) the second term on the right-hand side are the capillary volume forces depending on the gradient of the liquid density ρ^L and the interaction effects between solid and liquid as well as between liquid and gas manifested by $\hat{t}^{SL}$ and $\hat{t}^{GL}$. The last term of the right-hand side describes the friction volume force between liquid and solid as well as between gas and solid, where S_L, S_G are scalar response parameters. The corresponding friction forces between liquid nd gas can be neglected as a result of their small effects.

4 The Fundamental Field Equations of Poroelastodynamics

The fundamental system of field equations describing the motion of a saturated elastic porous body consist of the
strain-displacement relation (2.25)

$$\mathbf{E}_S = \frac{1}{2}(\nabla \mathbf{u}_S + \nabla^T \mathbf{u}_S),$$

$$e_S = \mathbf{E}_S \cdot \mathbf{I} = \text{div } \mathbf{u}_S, \quad (4.1)$$

the *density-displacement relations* (2.60)

$$\text{grad } \rho^L = -\rho_{0L}^L (\nabla \nabla \mathbf{u}_L)^{\overset{13}{T}} \mathbf{I}, \quad (4.2)$$

the *equation of motion for the partial phases* (2.66)

$$\text{div } \mathbf{T}^\alpha + \rho^L \mathbf{b} + \hat{\mathbf{p}}^\alpha = \rho^\alpha (\mathbf{u}_\alpha)''_\alpha, \quad (4.3)$$

where we have assumed a common external acceleration $\mathbf{b}$, and the *constitutive relations* (3.29) and (3.31) to (3.37)

$$\mathbf{T}^S = -n^S p\,(1 - B^S)\mathbf{I} + 2\mu^S \mathbf{E}_S + \hat{\lambda}^S e_S \mathbf{I}, \quad (4.4)$$

$$\mathbf{T}^L = -n^L p \quad (4.5)$$

$$\mathbf{T}^G = -n^G p\,\mathbf{I} + \mathbf{p}_E^G \mathbf{I}, \quad (4.6)$$

$$\hat{\mathbf{p}}^S = p \operatorname{grad} n^S - \hat{t}^{SL} \operatorname{grad} \rho^L - S_L[(\mathbf{u}_L)'_L - (\mathbf{u}_S)'_S]$$
$$+ S_G[(\mathbf{u}_G)'_G - (\mathbf{u}_S)'_S], \quad (4.7)$$

$$\hat{\mathbf{p}}^L = p \operatorname{grad} n^L + (\hat{t}^{SL} + \hat{t}^{LG}) \operatorname{grad} \rho^L + S_L[(\mathbf{u}_L)'_L - (\mathbf{u}_S)'_S], \quad (4.8)$$

$$\hat{\mathbf{p}}^G = p \operatorname{grad} n^G - \hat{t}^{LG} \operatorname{grad} \rho^L - S_G[(\mathbf{u}_G)'_G - (\mathbf{u}_S)'_S], \quad (4.9)$$

where the different velocities and acceleration have been replaced with the time derivatives of the displacement vectors $\mathbf{u}_\alpha$

We assume that the porous solid is homogenous and isotropic; therefore the material response parameters are not dependent on the position $\mathbf{x}$. With the calculation rules

$$\operatorname{div}(\mu^S \nabla \mathbf{u}_S) = \mu^S \triangle \mathbf{u}_S, \quad (4.10)$$

$$\operatorname{div}(\mu^S \nabla^T \mathbf{u}_S) = \mu^S \nabla \operatorname{div} \mathbf{u}_S, \quad (4.11)$$

$$\operatorname{div}[\hat{\lambda}^S \nabla (\operatorname{div} \mathbf{u}_S) \mathbf{I}] = \hat{\lambda}^S \triangle \operatorname{div} \mathbf{u}_S, \quad (4.12)$$

$$\nabla \operatorname{div}[n^S p (1 - B^S) \mathbf{I}] = n^S (1 - B^S) \nabla p \quad (4.13)$$

wherein $\triangle(...)$ denotes the Laplace operator, the fundamental field equations take the following forms inserting the constitutiv equations (4.4) through (4.9) into the equations of motion (4.3):

for the porous solid

$$(\hat{\lambda}^S + \mu^S) \triangle \mathbf{u}_S + \mu^S \operatorname{div} \operatorname{grad} \mathbf{u}_S - n^S (1 - B^S) \nabla p +$$
$$+ B^S p \nabla n^S - \hat{t}^{SL} \rho_{0L}^L (\overset{13}{\nabla} \mathbf{u}_L)^T \mathbf{I} - S_L[(\mathbf{u}_S)'_S - (\mathbf{u}_L)'_L] -$$
$$- S_G[(\mathbf{u}_S)'_S - (\mathbf{u}_G)'_G] + \rho^S \mathbf{b} = \rho^S (\mathbf{u}_S)''_S, \quad (4.14)$$

for the liquid

$$-n^L \nabla p + (\hat{t}^{SL} + \hat{t}^{LG})\rho_{0L}^L (\nabla\nabla \mathbf{u}_L)\overset{13}{\vphantom{I}^T}\mathbf{I} - S_L[(\mathbf{u}_L)'_L - (\mathbf{u}_S)'_S] +$$
$$+ \rho^L \mathbf{b} = \rho^L (\mathbf{u}_L)''_L, \quad (4.15)$$

for the gas

$$-n^G \nabla p + \nabla p_E^G - \hat{t}^{LG}\rho_{0L}^L (\nabla\nabla \mathbf{u}_L)\overset{13}{\vphantom{I}^T}\mathbf{I} - S_G[(\mathbf{u}_G)'_G - (\mathbf{u}_S)'_S] +$$
$$+ \rho^G \mathbf{b} = \rho^G (\mathbf{u}_G)''_G. \quad (4.16)$$

In addition, the saturation constitution has the differential form

$$(n^S)'_S + (n^L)'_L + (n^G)'_G - \nabla n^L \cdot [(\mathbf{u}_L)'_L - (\mathbf{u}_S)'_S] -$$
$$- \nabla n^G \cdot [(\mathbf{u}_G)'_G - (\mathbf{u}_S)'_S] = 0. \quad (4.17)$$

The set of fundamental equations (4.14) to (4.17) constitute the governing equations of the boundary and initial value problem for liquid and gas filled linear elastic porous solids if appropriate boundary conditions and initial conditions are presented. The system of the fundamental equations represents a counterpart to the linear elasticity theory of one-component elastic bodies. The equations serve particularly as the basic equations for numerical investigations. From the above set of equations the weak formulations of dynamics within the framework of the finite element method can easily be derived and can be introduced into existing computer programs. Eqn. (4.14),(4.15), (4.16), and (4.17) turn over to the corresponding fundamental equations, of a binary incompressible elastic, non-capillary porous medium, already widely used (see, e.g, Liu and de Boer, 1999)

$$(\lambda^S + \mu^S)\triangle \mathbf{u}_S - \mu^S \operatorname{div} \nabla \mathbf{u}_S - n^S \nabla p -$$
$$- S_L[(\mathbf{u}_S)'_S - (\mathbf{u}_L)'_L] + \rho^S \mathbf{b} = \rho^S (\mathbf{u}_S)''_S \quad (4.18)$$

$$-n^L \nabla p - S_L[(\mathbf{u}_L)'_L - (\mathbf{u}_S)'_S] + \rho^L \mathbf{b} = \rho^L (\mathbf{u}_L)''_L \quad (4.19)$$

$$\text{div}\left(n^S \frac{\partial \mathbf{u}_S}{\partial t} + n^L \frac{\partial \mathbf{u}_L}{\partial t}\right) = 0 \quad (4.20)$$

where λ_S and μ_S are the Lamé constants setting $K^{SR} \to \infty$ in (3.30) and where the convective terms in (4.17) have been neglected.

These fundamental equations (4.18) through (4.20) have already been successfully applied to the investigation of dynamics problems (see, e.g, Liu and de Boer, 1999, and the cited references).

5 Conclusions

In this article the mechanical behavior of a ternary capillary-porous medium model with compressible elastic solid, incompressible liquid, and compressible gas phases have been treated within the framework of the geometrically linear theory and the fundamental equations, formulated in the displacements, have been developed. The set of the fundamental equations contains the incompressible binary model as a special case investigated in the past.

The new ternary model will be the basis for the derivation of general theorems in poroelasticity as the proof of uniqueness and maximum and minimum problems, which are convenient in the numerical (finite element method) treatment of complicated boundary and initial value problems. Moreover, it will be used to solve analytical and numerical dynamic problem as, e.g., wave propagations, as has been done for a binary, incompressible porous medium. The derivation of general theorems and the solution of dynamic problems will be reported elsewhere.

References

1925 von Terzaghi, K.: Principles of soil mechanics, *Engineering News-Record* **19**, 742–746, **20**, 796–800, **21**, 832–936, **22**, 974–978, **23**, 912–915, **25**, 987–999, **26**, 1026–1029, **27**, 1064–1068.

1957 Biot, M. A.; Willis, D. G.: The elastic coefficients of the theory of consolidation, *Journal of Applied Mechanics* (J App Mech) **24**, 594–601.

  a Truesdell, C.: Zur Geschichte des Begriffs innerer Druck, *Physikalische Blätter* (Phys Bl) **12**, 315–326.

b Truesdell, C.: Das ungelöste Hauptproblem der endlichen Elastizitätstheorie, *Zeitschrift für angewandte Mathematik und Mechanik* (ZAMM) **36**, 97–103.

1960 Truesdell, C.; Toupin, R.: The Classical Field Theories, in: Handbuch der Physik (ed. S. Flügge) Vol. III/1, Springer-Verlag Berlin • Göttingen • Heidelberg.

1964 Kelly, P. D.: A reacting continuum, *International Journal of Engineering Science* (Int J Eng Sci) **2**, 129–153.

1965 Eringen, A. C.; Ingram, J. D.: A continuum theory of chemically reacting media I, *International Journal of Engineering Science* (Int J Eng Sci) **3**, 231-241.

Green, A. E.; Naghdi, P. M.: A dynamical theory of interacting continua, *International Journal of Engineering Science* (Int J Eng Sci) **3**, 231–241.

1967 Mills, N.: On a theory of multi-component mixtures, *Quarterly Journal of Mechanics and Applied Mathematics* (Quart J Mech appl Math) **20**, 449–508.

1968 Müller, I.: A thermodynamic theory of mixtures of fluids, *Archive for Rational Mechanics and Analysis* (Arch Rat Mech An) **28**, 1–39.

1969 Bowen, R. M.; Wiese, J. C.: Diffusion in mixtures of elastic materials, *International Journal of Engineering Science* (Int J Eng Sci) **7**, 689–722.

Truesdell, C.: Rückwirkungen der Geschichte der Mechanik auf die moderne Forschung, *Humanismus und Technik* **13**, 12–13.

1980 Bowen, R. M.: Incompressible porous media models by use of the theory of mixtures, *International Journal of Engineering Science* (Int J Eng Sci) **18**, 1129–1148.

1984 Truesdell, C.: Rational Thermodynamics, second ed., Springer, New York • Berlin • Tokyo.

1986 de Boer, R.; Ehlers, W.: Theorie der Mehrkomponentenkontinua mit Anwendung auf bodenmechanische Probleme, Teil I, Forschungsberichte aus dem Fachbereich Bauwesen der Universität-GH Essen, Heft 40, Essen.

1996 de Boer, R.: Highlights in the historical development of the porous media theory – toward a consistent macroscopic theory, *Appl Mech Rev* **49**, 201-262.

1997 de Boer, R.; Didwania, AK.: The effect of uplift in liquid-saturated porous solids – Karl Terzaghi's contributions and recent findings, *Géotechnique* **47**, 289-298.

Bluhm, J.: A consistent model for saturated and empty porous media, Forschungsberichte aus dem Fachbereich Bauwesen **74**, Universität-GH Essen.

Bluhm, J.; de Boer, R.: The volume fraction concept in the porous media theory, *Zeitschrift für angewandte Mathematik und Mechanik* (ZAMM) **77**, 563–577.

1999 Ehlers, W.; Volk, W.: Localization phenomena in liquid-saturated and empty porous solids, *Transport Porous Media* **34**, 159-177.

Liu, Z.; de Boer, R.: Propagation and evolution of wave fronts in two-phase porous media, *Transport Porous Media* **34**, 249-267.

2000 de Boer, R.: Theory of Porous Media: Highlights in the historical development and current state. Springer-Verlag Berlin • Heidelberg.

2002 de Boer, R.; Didwania, AK.: Capillarity in porous bodies: Contribution of the Vienna school and recent findings, *Acta Mechanica* **159**, 173-188.

From Hilbert space to the number of Higgs particles via the quantum two-slit experiment

by Mohamed S. El Naschie

Abstract

Rigorous mathematical formulation of quantum mechanics requires the introduction of a Hilbert space. By contrast, the Cantorian E-infinity approach to quantum physics was developed largely without any direct reference to the afore mentioned mathematical spaces. In the present work we present a novel reinterpretation of basic $\varepsilon^{(\infty)}$ Cantorian spacetime relations in terms of the Hilbert space of quantum mechanics. In this way, we gain a better understanding of the physical and mathematical structure of quantum spacetime. In particular we show that the two-slit experiment required a definite topology which is consistent with a certain fuzzy Kähler manifold or more generally a Cantorian spacetime manifold. Finally by determining the Euler class of this manifold, we can estimate the most likely number of Higgs particles which may be discovered.

1 Introduction

In numerous previous publications, one of the most fundamental experiments in quantum physics, namely the two-slit experiment, was reconsidered [1-5]. The main philosophy of our derivation was to imagine ourselves going back in time before quantum mechanics was invented but after T. Young proved the wave character of the light. We suppose furthermore, that our technical possibilities were expanded to the state where the light source could be made so weak that only one photon at a time could be released [1-5]. In other words, we have a manifest particle-wave paradox but no quantum theory. However we have probability theory as well as a lesson taught to us by special and general relativity. This lesson could be summarized into two important statements. First, spacetime geometry and topology affects gravity and thus, physics. Second, the form of spacetime as it appears to us does not necessarily need to be the real form. Relativity clearly showed us that the $3+1$ spacetime of our observation which is not really the $4D$ spacetime of photons and fast electrons. In a similar way, the flat spacetime of classical mechanics is only an approximation to real spacetime in the large which is curved [6,7].

Our resolution to the problem starts from a heuristic point of departure and arrives at an utterly simple equation [1-5]

$$|P_1 \pm P_2| = |\pm P_1 P_2| \qquad (1)$$

where P_1 is the probability for a photon to pass through slit number one and P_2 is the probability that it will pass through slit number two.

The decisive step comes however when the above is interpreted as a statement about the very structure of quantum spacetime. In particular it was shown that the above equation implies that we have two and seemingly only two ways to understand what the two-slit experiment is saying. The first possibility is to take spacetime at its face value as a classical $3 + 1$ flat space or in the relativistic case, as a $4D$ Mikuwiski spacetime but then we would have to accept probability with phase and complex wave function. Alternatively, we insist on real probabilities and end up accepting that real quantum spacetime is a fuzzy Kähler-like manifold which we call $K(\varepsilon^{(\infty)})$ or fuzzy K_3 [3]. This fuzziness of this $K3$-like manifold was shown elsewhere to relate to the golden mean probabilities and an Euler class $\chi = 26 + \phi^3$ instead of $\chi = 24$ where $\phi = (\sqrt{5} - 1)/2$ is the golden mean [1-5].

The above conclusion brings our theory very close to modern theories of super strings which although very popular and rightly so at present, they could not be seen as being in line with the classical form of quantum mechanics and quantum field theory with its Hilbert space and Fock space respectively [8,9]. However we feel that a great deal could be gained by interpreting the above result as well as E-infinity theorem [6,7] in terms as close as possible to Hilbert space and conventional quantum field theory with its Fock space extension of the Hilbert space of orthodox quantum mechanics [8,9]. Such an attempt is worthwhile because at a minimum a clearer picture of both approaches will arise. In fact, we will show how we may use these results to estimate the number of Higgs particles which we could discover experimentally in the foreseeable future.

2 Hilbert space

A mathematical text book definition of Hilbert space is the following [10,11].

Definition: A Hilbert space is an inner product space which is a complete metric space with respect to the metric induced by its inner product.

A more general form of Hilbert space is a Banach space. In fact, every Hilbert space is a Banach space, but in applications it is the special class of Hilbert spaces which are more important particularly with respect to quantum mechanics. There we see that the space of bra or ket vectors, to use Dirac's terminology, is a Hilbert space as long as the vectors are restricted to be of finite length and to have finite scalar products. In general however, bra and ket vectors used in quantum mechanics form a more general space than a Hilbert space [9].

Let us be more concrete with regard to our quantum application of Hilbert space. Consider the spin of an electron. In such a case we have only two states; spin up and spin down. This would be a two dimensional state vector space. However we also have quantum situations where the number of possibilities is unbounded. In such a case, one must use infinite dimensional spaces to represent them. This is actually what one usually associates with a Hilbert space, so that we may give another definition.

Definition: Hilbert space is a mathematical theory of vectors in a space of infinite dimensions.

From the above it follows that Hilbert space is a particular form of vector space. In addition we note a similarity with $\varepsilon(\infty)$ spacetime due to the fact that both spaces live in infinite dimensions, but for the moment that is all, because Hilbert space is supposed to be an abstract mathematical construction while $\varepsilon(\infty)$ is supposed to model real spacetime at all possible resolutions including the quantum. Clearly as we move from quantum mechanics to quantum field theory, we do move from Hilbert space to a Fock space [8-11].

3 The master equation of $\varepsilon(\infty)$ Cantorian spacetime and Hilbert space

One of the main equations of E-infinity theory is the master equation which fixes the expectation value of the dimensionality of $\varepsilon(\infty)$, namely [6,7]

$$-<n> = \sum_{n=1}^{\infty} (\phi^n)(n) \qquad (2)$$

Now let us set $-<n> = <\hat{A}>$, $n = \lambda_n$ and $\phi_n = a_n$. That way one finds [11]

$$<\hat{A}> = \sum_{n=1}^{\infty} a_n \lambda_n \qquad (3)$$

Clearly this is the well known quantum mechanical formula for the expectation value $<\hat{A}>$ of an observable operator $\hat{A}$ in the state $\Psi(x)$ of a physical system defined by [11]

$$<\hat{A}> = \frac{<\Psi|\hat{A}\Psi>}{<\Psi|\Psi>} \qquad (4)$$

The value a_n is the weight of the eigen values λ_n. Consequently there is a complete analogy between the mathematical foundation of quantum mechanics and the foundation of E-infinity theory. We recall that $\left(d_c^{(0)}\right)^n = \phi^n$ is indeed the weight of a dimension n in $\varepsilon(\infty)$ from which the well known $\varepsilon(\infty)$ result $<n> = 4 + \phi^3$ follows [6,7]. Seeing it in this way, Hilbert space starts to appear less of an abstract mathematical concept and far more related to the real spacetime of $\varepsilon(\infty)$. In the following we would like to pursue this view point as far as we can with the obvious aim of giving a more conventional interpretation to the undecidability condition [1-5] which we have established for the two-slit experiment and the fuzzy K_3 spacetime model [3].

4 The significance of Hilbert space

We explained earlier on that mathematically, Hilbert space is an inner product space. Its main physical significance is that it represents the wave function of the quantum system. This means it represents the system prior to measurement. When measurement is taken the system collapses into a specific state. The collapsed state is what is nearest to the classical concept of a point. We have also seen that there is a correspondence between $<n>$ and $<\hat{A}>$ and therefore the wave functions in $<\hat{A}>$ corresponds to a scalar [6,11]. Consequently the inner product degenerates to a simple multiplication. This in turn would correspond to the right hand side of our fundamental equation 1. We could look at the same point from a different perspective, namely that of dimension, taking into consideration that P_i may be interpreted as probabilities as well as dimensions [6,7]. In this way the distinction between a mathematical Hilbert space and a physical E-infinity spacetime is blurred. In the simplest of terms, one could say that the rule of jointing two classical spaces is to add their dimensions. This would correspond to the left hand side of our fundamental equation. By contrast, when joining two Hilbert spaces, we must multiply the dimensions. Thus our undecidability equation may be interpreted as an equation balancing the dimensionality of Hilbert space where the probability wave function is propagating, with the dimensionality of the measurement space where the real measurement is taken [1-7]. From a positivistic view point, Hilbert space as well as E-infinity spacetime are neither entirely mathematical nor entirely physical but they are an entirely useful way to model reality.

5 Interpretation via Töplitz operator

The theory of Hilbert space is related to the subject of infinite Töplitz and Hankel matrices and also to the index theorem of Töplitz [10]

$$Ind(T(f)) = <[\eta] \mid f^*[T]> \qquad (5)$$

The vital point is remarkably the one to one correspondence of the index theorem to our undecidability condition [1-5]. This is because while the left hand side gives an expression of the winding number in terms of an operator which may be seen as an analogue to the particle interpretation, the right hand side expresses in some way a property of waves related to the fact that the phase of f varies by a multiple of 2π as x varies by one period. Again we are de facto having a balancing act between wave and particle which is not possible in a classical spacetime geometry.

In other words, our solution of the two-slit experiment is generic and constitutes a genuine resolution of the wave particle duality via spacetime topology, without the need to invoke orthodox quantum mechanics. In fact the conclusion of Tanaka that E-infinity is at least occasionally simpler and superior to quantum mechanical calculation seems to have some justification in the present analysis [12].

6 The Higgs number from $K(\varepsilon(\infty))$ topology

In various recent publications we have shown that a logical resolution of the two-slit experiment is possible in a fully realistic manner once we give up our familiar notion of smooth $3+1$ spacetime in favour of the fuzzy K_3-like manifold of E-infinity theory [1-7]. It was also shown that the Euler class of this manifold is not $\chi = 24$ as K_3 nor $\chi = 26$ as M^4 but rather $\chi = 26 + \phi^3$ where ϕ is the golden mean. The next important step is the realization that within Donaldson Yang-Mills moduli spaces theory, the curvature density of an anti-self-dual connection of a bundle over a closed manifold may be represented as a topological invariant [14]

$$\int_x |F(A)|^2 = -\int_x Tr(F(A)^2) = 8\pi^2 K(E) \qquad (6)$$

where A is the connection, E is the bundle and x is the manifold. In other words, for the normalized case we have the curvature densities

$$|F(A)|^2 = 8\pi^2 \qquad (7)$$

and since the Chern-Simons-Weil theory tells us that $kC_2(E)$ we see that the curvature measures converge towards a limit θ and one finds [14]

$$\int_x d\nu = 8\pi^2 k \tag{8}$$

In the case of E-infinity theory the integral is replaced by an infinite sum in the usual way and consequently we have [1-7]

$$\int_x \longrightarrow \sum_0^\infty \tag{9}$$

$$8\pi^2 \longrightarrow 2(42 + \phi^3) = (20)(<n>) \tag{10}$$

$$k \longrightarrow \chi = 26 + \phi^3 \tag{11}$$

Consequently the curvature density of the fuzzy $K(\varepsilon(\infty))$ is given by

$$2(42 + \phi^3)(26 + \phi^3) = (84.72135954)(26 + \phi^3) \tag{12}$$

$$= 2218.033989 \tag{13}$$

To bring this value down to $D = 3 + 1$ we have to realize that it corresponds to $D = (3+1)(D^{(8)}) = 32$ so that it must be divided by 32 which gives us

$$\frac{2218.033989}{32.3606977} = \bar{\alpha}_0/2 \tag{14}$$

$$= 68.54101965 \tag{15}$$

Noting that the standard model as it stands today contains 60 experimentally verified elementary particles, we see that we are missing some 9 particles as recently discussed by He [15]. Excluding the graviton one would be tempted to say that we may expect 8 Higgs-like particles to be discovered in the foreseeable future. However taking other considerations into account, it seems that it is more likely that these 8 elementary particles are degrees of freedom and we should be expecting either one or five Higgs particles as reasoned in earlier publications [1-6]. The important point here is that we started with an experiment, namely the two-slit experiment with quantum particles and proceeded from there using theoretical reasoning to the point where we could make theoretical predictions about what we could expect to find experimentally. This procedure seems to lie well within the best classical tradition of theoretical physics.

7 Conclusion

There is clearly an analogy between quantities defined in Hilbert space and those representing $\varepsilon(\infty)$ Cantorian spacetime. For instance in Hilbert space we have [11]

$$<\hat{A}> = \int_{n=1}^{\infty} a_n \lambda_n \qquad (16)$$

while in $\varepsilon(\infty)$ we have [6]

$$<n> = \int_{n=1}^{\infty} \phi^n n \qquad (17)$$

In addition the undecidability condition of the two-slit experiment [4] of Equation 1 turns out to express the same fact as the Töplitz index theorem as well as the dimensional union in classical and Hilbert spaces. Our general conclusion is that the particle-waves duality can find a semi-classical resolution if we assume that micro spacetime is a fuzzy Kähler-like manifold akin to $\varepsilon(\infty)$. Therefore if we accept spacetime as classical, then we must also accept the wave-particle duality. Alternatively we could insist on classical probabilities without a phase, then we have to take the real complex character of $\varepsilon(\infty)$ micro spacetime into consideration.

We may mention that the connection between chaotic quantum spacetime geometry and topology and the paradoxal outcome of the quantum two-slit experiment has been recognized for a long time and illustrated by the analogy to the three points and four points chaos game [13]. Thus from a purely positivistic view point, it is irrelevant to ask if Hilbert space of E-infinity spacetime is real or a mathematical construction. What is important is that they are useful and transmit a feeling of understanding. In the case of Hilbert space the mathematics was perfect but this feeling of understanding was never complete. With the present analogy between Hilbert space and $\varepsilon(\infty)$ the situation is much better. In addition to that the power of predictability has increased enormously when we go as far as making precise quantitative statements for the number of Higgs particles which one could expect to discover experimentally. Thus starting from experimental facts, we proceed to determine the topology and geometry which could make sense out of these facts. Subsequently by analysing the corresponding non-classical spacetime manifold we were able to make a statement about the elementary particles content of these manifolds. That way we are back to experimental evidence from which we expect a confirmation of the predictions of the theory, namely either one or five Higgs particles as well as a Higgs field with eight degrees of freedom in addition to one graviton. It would be an incredible triumph of the geometrical topological method in physics if some of the preceding conclusions could be verified experimentally.

In conclusion I would like to say a few words of thanks to all those wonderful friends who gathered here today to, dare I say, console me about reaching this age which only three decades ago seemed so absurdly old to me. I was only twenty years old when we used to say never trust a man older than thirty! I am now baffled to say that although the body is no longer so willing, the spirit is still burning with desire to know more and to go on intellectual adventures into unchartered seas. I thank from the bottom of my heart for the hard work of Otto Rössler, Hans Diebner, Peter Weibel and Garnet Ord who made this memorable gathering possible. Thanks also to Paulo Grigolini and Gerardo Iovane, my dear Italian friends for their valuable contributions. Of course I miss the presence of my very dear friends Itamar Procaccia and S. Al Athel as well as Gamal Al Ghitany, Adel Hussein and Galal Abdel Hamed Abdulla who could not attend. In the Moslem theology, man is nothing but a memory and this wonderful day reminds me of a memorable trip to Paris when I was sixteen years old where I came across a wonderful sentence by the enigma of German romantic poetry Jean Paul which summarizes it all and at the risk of appearing too sentimental (which I am), I would like to quote it "Memories are a paradise from which man could not be evicted". To all who have helped in making these memories I say thank you my dear, dear friends.

References

[1] M.S. El Naschie. On a fuzzy Kähler-like manifold which is consistent with the two-slit experiment. *Int. Journal of Nonlinear Science & Numerical Simulation*, 6(2):95–98, 2005.

[2] M.S. El Naschie. Nonlinear dynamics of the two-slit experiment with quantum particles. *Int. J. Problems of Nonlinear Analysis in Engineering Systems*, Kazan University, Russia (In press).

[3] M.S. El Naschie. From experimental quantum optics to quantum gravity via a fuzzy Kähler manifold *Chaos, Solitons & Fractals*, 2005 (In press.)

[4] M.S. El Naschie. Non-Euclidean spacetime structure and the two-slit experiment. *Chaos, Solitons & Fractals*, 2005 (In press).

[5] M.S. El Naschie. A new solution for the two-slit experiment. *Chaos, Solitons & Fractals*, 2005 (In press).

[6] M.S. El Naschie. A review of E-infinity theory and the mass spectrum of high energy particle physics *Chaos, Solitons & Fractals*, 19(1):209–236, 2004.

[7] M.S. El Naschie. Emerging research fronts. Comments by Mohamed El Naschie *ISI Essential Science Indicators*. http://esi-topics.com/erf/2004/October04.MohamedElNaschie.html.

[8] M. Kaku. Quantum Field Theory, Oxford, 1993.

[9] P.A.M. Dirac. The principles of quantum mechanics. Oxford, 1987.

[10] N. Young. An introduction to Hilbert space. Cambridge, 2004.

[11] L. Debnath and P. Mikusinski. Hilbert spaces with applications. Academic Press, London, 1999.

[12] Y. Tanaka. The mass spectrum and E-infinity theory. *Chaos, Solitons & Fractals*, 2005 (In press).

[13] M.S. El Naschie. Iterated function systems, information and the two-slit experiment of quantum mechanics. In: Quantum Mechanics, Diffusion and Chaotic Fractals. Edited by M. El Naschie, O. Rössler and I. Prigogine. Elsevier-Pergamon Press, Oxford, p. 185-189, 1995.

[14] S.K. Donaldson and P.B. Kronheimer. The geometry of four-manifolds. Oxford, 1990.

[15] Ji-Huan He. In search of 9 hidden particles. *Int. J. Nonlinear Science & Simulation*, 6(2):95–98, 2005.

Springer-Verlag and the Environment

WE AT SPRINGER-VERLAG FIRMLY BELIEVE THAT AN international science publisher has a special obligation to the environment, and our corporate policies consistently reflect this conviction.

WE ALSO EXPECT OUR BUSINESS PARTNERS – PRINTERS, paper mills, packaging manufacturers, etc. – to commit themselves to using environmentally friendly materials and production processes.

THE PAPER IN THIS BOOK IS MADE FROM NO-CHLORINE pulp and is acid free, in conformance with international standards for paper permanency.